THE HIDDEN GIANTS

Women Hold Up Half the Sky

4,000 YEARS OF WOMEN IN SCIENCE AND TECHNOLOGY

ISBN

978-0-615-73267-1

Dedicated to Charcoal

PREFACE TO THE THIRD EDITION

Science is a traditional role for women. For over 4,000 years of written history women have participated in this great human adventure. Science and technology are neither new nor difficult for women any more than they are for men. Yet the stories of many of our scientists do not form part of our instruction in science from kindergarten through college. Missing from our textbooks and data are the fundamental contributions of scientists, both male and female but especially female. Female creativity and genius spill out of our technical past. The stories of these women not only provide role models for future scientists, but they also strengthen and broaden our ability to deal with the present. An Internet site devoted to the participation and success of women in the technical history of humanity accompanies this book. This site is used by school systems world wide as a student resource.

http://www.astronomy.ua.edu/4000WS

This book contains material on the history of women in science and technology and the wonderful work they have done down through the centuries. The intended audiences are the interested public, college students, and teachers.

The 3rd edition is published under the imprint of the Washington Academy of Sciences.

Sethanne Howard, PhD

December 2012

The Washington Academy of Sciences incorporated in 1898 as an affiliation of eight Washington D.C. area scientific societies. The founders included Alexander Graham Bell and Samuel Langley, Secretary of the Smithsonian Institution. Then, as now, the purpose of the new Academy is to encourage the advancement of science and "to conduct, endow, or assist investigation in any department of science."

The Academy offers Academy members who have written a science-heavy book and intend to self-publish the opportunity to submit the book to our editors for review of the science therein. The manuscript receives the same rigorous scientific review that we accord articles published in our Journal. If the reviewer(s) determine(s) that the science is accurate, the author may then continue with the self-publishing process, and the book will be issued under the imprint of the Washington Academy of Sciences. *The Hidden Giants*, 3rd edition, has met that criterion.

TABLE OF CONTENTS

CHAPTER 1

THE BEGINNINGS

What I have done here no one has done before[1]

    Calling all the hidden giants of science and technology – the ones left out of history books! Seek them in the dusty halls of archeology and myth and in fragments of rocks. The rewards are many. Once discovered, the stories of these hidden giants will refresh and delight you with their ingenuity, genius, and sheer stubbornness. We need to get them back into the history books, back into the center of inquiry, so we can draw on their strengths.

Dr. Gerda Lerner explained this challenge in her address as the incoming president of the Organization of American Historians.[2]

"...*All women have in common that their history comes to them refracted through the lens of men's observations and refracted again through a male-centered value system.... From that time on [the beginning of written history] women were educationally deprived and did not significantly participate in the creation of the symbol system by which the world was explained and ordered. Women did not name themselves; they did not, after the Neolithic era, name gods or shape them in their image.... If the bringing of women — half the human race — into the center of historical inquiry poses a formidable challenge to historical scholarship, it also offers sustaining energy and a source of strength.*"

What an exciting challenge it is.

The word 'science' actually comes from the Latin word for knowledge, *scientia*. The word 'scientist' – meaning a person who studies science – was apparently coined in 1840 CE by the British Association for the Advancement of Science.[3] Before the 19th century such people were usually called natural philosophers, scholars, technicians, and even amateurs.

'Technology' comes from the Greek word τεχνολογια (*technologia*) meaning the practical application of knowledge. This includes engineering.

'Mathematics' comes from the Greek work μάθημα (*máthēma*) meaning knowledge, study, and learning.

A scientist/technician is someone who studies the natural world, how it operates, moves, and changes.

And there are a lot of them. The number of United States workers in science and engineering occupations grew from about 182,000 in 1950 to 17 million in 2007 working in these broadly defined fields:

- ~ 10% Computer/Math (40% women, 60% men)
- ~ 10% Astronomy, Physics, Chemistry (25% women, 75% men)
- ~ 20% Engineering (10% women, 90% men)
- ~ 20% Biology/Life Sciences (50% women, 50% men)
- ~ 40% Social Sciences (50% women, 50% men)

where the first number on each line represents the percentage of scientists/technicians in that field of study. The percentages of women and men, however, in each field tell a particular story. Women have made great strides in the 'softer' sciences and lag seriously behind in the 'harder' ones.[4] In the first decade of the 21st century the number of students going into computer science has dropped significantly.

The notion that technical excellence is not for girls (or minorities) persists. Why? After all, "women hold up half the sky" is a saying native to many cultures. Women are half the human race. Why are we neglecting the talents of half our species? It is vital that we know what the women have done, how they have contributed. If we learn about those women who did succeed despite the odds, we might provide diverse and multi-faceted role models for young women and men today and thereby even out the distribution. Science and technology (natural philosophy) are innately diverse. We need role models that highlight and celebrate this diversity. When the role models are plentiful then that former university president[5] will know better than to state that innate differences in the brain limit women in science and technology. By getting those hidden giants back into the history books we can draw on their strength as much as we draw upon the strength of the other half of the human race – the males.

We all know that science[6] is an adventure, a trip that uncovers beauty with every new thing understood. *Everyone* deserves to share in this excitement and personal fulfillment. The results of science have no gender.

That is worth repeating. The results of science have no gender. We cannot back out of some invention, theory, or solution whether the originator was female or male. It is a *human* adventure.

Many things make us human, and one of them is our ability to affect and predict our environment. I call this **science** — the definition of structure for our world — **technology** — the use of structure in our world — and **mathematics** — the common language of structure — science, technology, and mathematics,[7] all have been part of our human progress, through every step of our path to the present. Women and men together have researched and solved each emerging need. Women and men together have defined the advancing path of this fundamental human activity. Women and men together have eased the burden for all of us. Women and men together have sought out this great joy – to be the first to see something new in the world. That excitement reaches the deepest part of our souls. To *think* is just about the greatest 'turn-on' that exists. It is irresistibly compelling. Most of the effort in science is repetitive, even boring. The excitement is exquisite and rare, and when it comes, it is the deepest joy and greatest wonder — all the labor is worth those few ecstatic moments. When I am the first to discover something new to the world, even a tiny something, it is the most exciting thing there is.

Understanding science will only strengthen our lives, our work, and our world. We want solutions to our problems. Solutions come from questions, research, thought, and technology. For as long as we have been human we have developed and thought about science. For as long as we have been human we have looked forward to the next challenge, the next goal, and the next creative thought.

Given that the results of science have no gender, what then are the attributes of the people who produce these results?

The attributes of the scientist are *intelligence* (the ability to combine information quickly, organize thoughts and coordinate actions to achieve results), *doubt* (the ability to question), *luck* (the ability to take quick advantage of an opportunity), *sweat* (the ability to work hard), and *courage* (the ability to maintain a clarity of thought despite opposition).

All scientists share these attributes. The scientist is often in the right place at the right time; *i.e.,* is lucky. The scientist absorbs as much education as possible. It is education that provides the grist for the mind to use any luck it encounters. The scientist has a nimble and adaptable mind, well-equipped to doubt. And finally, a scientist works hard — very, very hard. Both women and men have these attributes. There is no gender lurking in this definition. Women have courage aplenty. Women share the common intelligence of humanity. They are superlative doubters. The sweat of their bodies waters all the monuments of the world. Many have shared luck with their male brethren. We need to celebrate these women along with the men and raise them all to be heroes.

And so we ask: where are the women? *Are* there just a few women of scholarship?

The answer is a resounding no! There are many, many thousands, even millions in the United States alone. But the women are mostly missing from the written record. When we look in the history books we rarely find them. When asked to name a woman of science, the typical person can think of only one – Marie Curie.

Of course very few *people* in history were literate and numerate, so one finds just a smattering of scholarly men in the records, and I suspect that the even rarer scholarly women were the lucky ones who had access to study and freedom from the drudgery of the general wifely state which was completely occupied with the skills of housekeeping. Access to scholars and information has always depended upon gender, location, birth, and luck. If one was born to a secure family then one might learn to read, write, and cipher. Men had the advantage here. Therefore, if a woman was literate and numerate, she was likely to have links to a tutor, a benevolent father, husband, or brother who was willing to share knowledge. Literacy was a privilege for both men and women, but especially for women. Perhaps, though, she lived during the Middle Ages when women had the great convent schools of England, France, and Germany open to them[8] or perhaps she lived at the dawn of history when she was honored.[9]

Regardless, the overwhelmingly vast majority of people, both male and female, had no access at all. They labored for their very food and shelter. For example, in England as late as 1841, 33% of men and 44% of women signed marriage certificates with their mark because they were unable to write. The freedom to specialize in scholarship rarely put food onto the table. This freedom springs from the human need to dream a future. Those who are freed to dream are freed by the willing labor of the rest. One of the greatest strengths of our species is its recognition that scholarship is worthy, is important, is valuable, and is necessary. The right to question must be sacrosanct. By recognizing that solutions to the very human need for food and shelter come from answering questions, humans met their needs.

The women who questioned did exist and were honored, even though their numbers seem to be few. It would be wrong to assume that women held no positions of scholarship just because there are so few listed in history books. These wonderful women existed and formed a crucial part of that thread of scholarship and invention that runs strong through history.

To bring these women out of obscurity and put them into the center of history and science is our goal – to turn the hidden giants into known friends. We have to look just about everywhere. Scholarship is the key word, not science. The word 'scientist' is rather new, as I said before; today it means someone with an advanced degree (often a Doctorate in Philosophy, a PhD) who works in a technical field. This person has studied a narrowly defined field of research and often is well trained in only that field.

But looking just for PhDs is not insufficient. We need to include toolmakers, inventors, physicians, nurses, poets, and most importantly natural philosophers. "Natural philosophers" are those whose endeavors typically covered the classic seven liberal arts — grammar, rhetoric, logic, arithmetic, geometry, music, and astronomy. Note that four of these are technical. A literate person perforce meant a numerate person[10]. So we look for our women in all fields of endeavor. We need to look outside of schools, because schools did not always exist, and because women could rarely take

7

advantage of those schools that did exist. When schools did not train scientists, learned people were either self- or privately taught.

So we look for holders of scholarly degrees from schools, yes, and also for poets and authors, architects and gardeners; we look in industry, in school lists, in textbooks, letters, and stories. The names of scholars may be deduced out of their poems, music, and writings. To track this, one needs to look at inventors and toolmakers as well as scholars.

Do we look for someone who changed the world? We can easily remember the very few people, both male and female, who produced something with a value that lives through centuries. These are the paradigm shifters — the rare geniuses, people like Sir Isaac Newton and Albert Einstein (and the ones like Sylvia Earle whom you will meet in these pages). Newton and Einstein defined the basic rules for the physical structure of our Universe. But they lived centuries apart from each other, one in the 17th century and one in the 20th century, thus indicating how rare such genius is. So do we look for just those few geniuses who changed the way the world thinks? No.

There are people, far, far greater in number than the paradigm shifters, who produced something of value for their time and place and possibly for many times and places. These people are much more difficult to find, and they are important. They provide the bedrock upon which the rare genius can build a new paradigm. As Sir Isaac Newton said "If I have seen farther than other men, it is because I stood on the shoulders of giants". Science is not an isolating activity; it is dependent upon the work of many. These women and men are important; they are special. They are the hidden giants.

To start the search let us consider three books from the 20th century. Mozan's book, *Woman in Science*, written in 1913[11] lists over 350 technical women of the past. This book is an amazing *tour de force* combining romantic views of women with solid references to original sources. Asimov's *Biographical Encyclopedia of Science and Technology,*[12] some 50 years later lists only sixteen women. Moore's book *Men of the Stars,*[13] a mere decade after Asimov's book, has none! This is a disappointing trend; however, the past decade has produced many publications about technical women. The 20th century is covered rather well (despite Moore's misogynist book). But it is misleading to assume that women were not scholars before the 20th century just because their names are missing from the history texts. Their absence is involuntary — a result of how history was compiled. We all have just opened the treasure box. These women contributed much. They had the entire universe to play with, to study and to enjoy. They were not left out of this great human experience. Let us start the search.

The Story Begins

In 400 CE[14] in Alexandria, Egypt there existed one of the world's rare treasures: the Great Library, founded by Ptolemy II (285-246 BCE) of Egypt. The Library was the center of scholarship for the entire Mediterranean world. Scholars came there to study and to teach. The Library grew in size and reputation through the gathering of the literary works until it became the collective brain of the Mediterranean. This was often rigorously enforced by the forceful taking of books from ships as they docked in the harbor and delivering them to the Library. The best minds of antiquity came there to study medicine, biology, astronomy, mathematics, literature, and geography. Critical editing was invented there. We build on those foundations still.

Typically, scholars could only bring texts *into* the Library. They could not remove the books themselves, only copies. The original texts of most of the Mediterranean world's literature were held there. Ptolemy III, Euergetes, (246 – 222 BCE) wished to borrow from Athens the original manuscripts or official state copies of the tragedies of Sophocles, Aeschylus, and Euripides. To the Athenians, these were cultural icons. They were reluctant to let the manuscripts out of their hands even for a moment. Only after Ptolemy guaranteed their return with an enormous cash deposit did they agree to lend the plays. But Ptolemy valued those scrolls more than gold or silver. He forfeited the deposit gladly and enshrined the originals in the Library. The outraged Athenians had to content themselves with the copies that Ptolemy presented to them. Rarely has a state so avidly supported the pursuit of knowledge. Tradition says the Library held upwards of half a million texts. Alexandria became the publishing capital of the western world, every book copied by hand. The Library was the repository of the most accurate copies in the western world.

As an example of its worth, the Old Testament of the Bible comes down to us mainly through the Greek translations made in the Library. Euclid lived and worked there. His book on geometry, the *Elements*, is perhaps the second most translated book after the Bible. A citizen of Alexandria was not only a citizen of a great city but also a citizen of the known world. The city encouraged and financed scientific research. The Library was eventually destroyed, not once but many times; however, from the time of its creation until its ultimate destruction it was the heart and brain of the ancient world. Many philosophers and teachers worked there, including the philosopher/mathematician **Hypatia** (He-pa-ti′-a) (c. 355 – 415 CE).[15] When she was 19 years old a mob of Christians tried to destroy the Library. It survived a little longer before its final destruction, ending an irreplaceable tradition of scholarship.

The glory of the Alexandrian Library is, unfortunately, a dim memory. The loss is incalculable. In some cases, we know only the titles of the works that were destroyed. We know that of the 123 plays of Sophocles in the Library, only seven survived. Similar survival numbers apply to the works of Aeschylus and Euripides. In most cases, we know neither the titles nor the authors.

Hypatia is not the earliest woman of science, technology, and invention. We shall meet that lady a bit later. Hypatia is, perhaps, one of the more well known which is why I start with her. During the time of Hypatia Alexandria was still a vital center of scholarship.

Hypatia was an astronomer, mathematician, philosopher, and head of a school of neo-Platonic philosophy, an amazing suite of talents for anyone. She wrote at least three books, none of which survive: a treatise on the *Conics of Apollonius,* a *Commentary on the Arithmetic of Deophantus*, and an *Astronomical Canon.* The first two were expositions of rather difficult mathematics, the third probably an exposition of planetary positions. It may be that a fourth commentary of hers did survive – her commentary on Ptolemy's *Handy Table,*[16] although this is uncertain. It may be that surviving editions of Ptolemy had been arranged and documented by Hypatia.

The little we know about her comes from letters sent to others. She designed many tools of her trade including an astrolabe[17] and a method for distilling water. She corresponded with people all over the Mediterranean, and letters addressed to 'the Philosopher' were delivered to her. She may have studied in

Athens at the neo-Platonic school conducted by Plutarch the Younger and his daughter Asclepigenia, although this is speculation.[18] Officials who assumed public responsibility in the city would call upon her as the leading philosopher in the city. Students flocked to her from all directions. Much was made of her beauty and eloquence. She wore the modest tribon — a coarse workaday garment worn by the poor, ascetics, and philosophers. Her words tell us something about her. They still echo with power:[19]

"All formal dogmatic religions are fallacious and must never be accepted by self-respecting person as final."

"Fables should be taught as fables, myths as myths, and miracles as poetic fancies. To teach superstitions as truths is a most terrible thing. The child mind accepts and believes them, and only through great pain and perhaps tragedy can he be in after years relieved of them. In fact men will fight for a superstition quite as quickly as for a living truth – often more so, since a superstition is so intangible you cannot get at it to refute it, but truth is a point of view, and so is changeable."

These words and others like them apparently led to her eventual downfall. In her time Alexandria was controlled by Rome; however, the growing Christian church was consolidating its power and wanted to eradicate pagan influence. Hypatia stood at the center of the grave forces warring in Alexandria, and her presence sparked the anger of the Bishop of the city – Cyril. Tradition tells us that a mob excited by Cyril dragged her from her chariot one day and armed with broken bits of pottery peeled her skin from her bones, scattering them to the winds and then burning her body. Cyril was made a saint.

A tragic ending for a brilliant woman. Did her sacrifice end such persecution of women who dared to think? While her death occurred some 1500 years ago, I must note that meteorologist **Dr. Ginous Mahmoudi** was executed by firing squad on December 17th 1981 by the Iranian Revolutionary Guard for expressing her minority faith and working as a woman of science.

Despite the dangers there were, and still are, many women who try the noble calling of science.

Figure 1 shows the painting by Raphael called the School of Athens. Placing Plato and Aristotle in the center, he also put Hypatia in the painting as the sole woman. She is indicated by the arrow.

Hypatia also said:

"Reserve your right to think, for even to think wrongly is better than not to think at all".

The history of women who dared to think becomes the history of the hidden giants. Where do we start?

We start with the world's earliest literature. The name of an appropriate woman appears there — over 4,000 years ago.[20] Science has been the business of women ever since. Certainly, though, women were questioners and thinkers long before that. Most myths, religions, and history place the beginnings of agriculture, laws, civilization, mathematics, calendars, time keeping, and medicine into the hands of women. And the mythology is so very rich. The stories form our common wealth. But whether it was the Goddess of Wisdom or War or Love (Figure 2) she was lost to the historical record yet kept strong in the dreams and myths of all peoples.

FIGURE 1 Raphael's Fresco the School at Athens with Hypatia as the only
female

FIGURE 2 Minoan Snake Goddess

The Western world owes much of its world view to a thick thread of scholarship that goes back to ancient Greece and earlier. Women belong in that thread. However, filtered through modern eyes, we still question whether women held high social status in that ancient world. Scarce records of the Near East put female gods at the heads of complicated religious pantheons, and so it is tempting to assume that this reflects a high status for women. It is a convenient conclusion but it represents only circumstantial evidence and is not proved. Eventually, though, the early Semite tribes wiped out the female dominated religions of the Near East replacing them with a variety of male gods. There are, however, a few things that hint at women having a respected status. The archeological record shows a predominance of female figurines from the Neolithic era. Females dominate in Minoan art.[21] Figure 2 shows the Minoan Snake Goddess found in the Minoan ruins (2700 to 1450 BC) – an example of the richness. The historical records are scanty and inferential at best; although, we can do a bit more than make a simple guess – see Chapter 4 for an outline of women's rights through the millennia. First let us see what the women themselves tell us.

Mesopotamia

It might be said that history begins in Sumer (in the lower flood plain of the Tigris and Euphrates rivers or 'Mesopotamia' – Greek for 'the land between the rivers' – modern day Iraq). While Egypt united under one pharaoh, and settlers from Mexico migrated to the Caribbean islands in dugout canoes, and western European farmers constructed large, stone chamber tombs, civilization developed in the lands around and between the Tigris and Euphrates rivers which flow southeast into the Persian Gulf. The earliest city-states developed on the fertile Mesopotamian Plain.[22] By 3500 BCE Uruk (or Ur) was probably the first substantial city-state. It covered over 1,000 acres in the land of Sumer (Figure 3). These early urban settlements were built around the central temple complex (later known as ziggurats) managed by a suite of priestesses and priests.

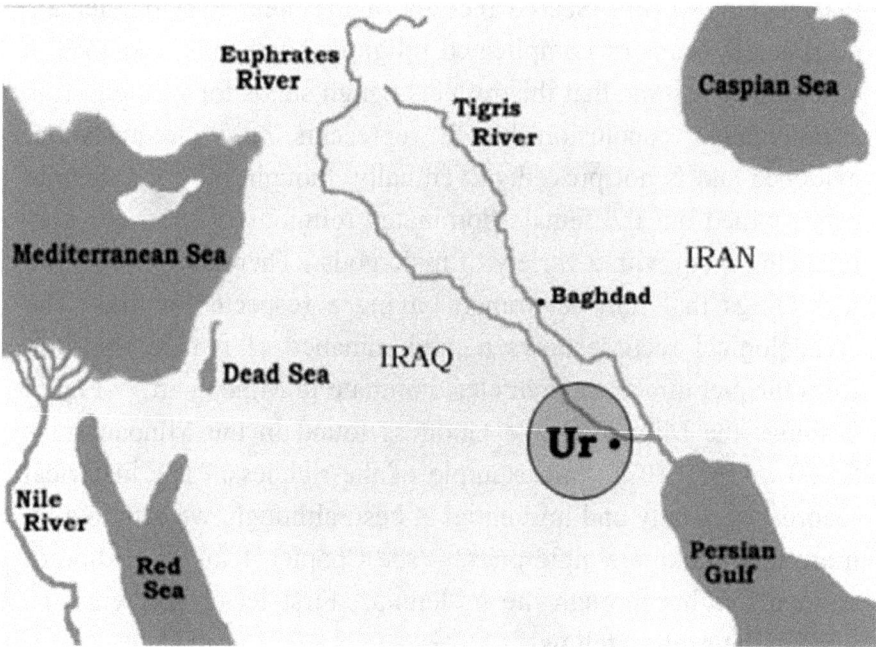

Figure 3 Location of Ur in southern Iraq

The Sumerians developed cuneiform writing, a way of arranging impressions stamped on clay by the wedge-like section of a chopped-off reed. Eventually the signs stamped onto the clay became more than mere pictographs. They came to stand for syllables and sounds. Our modern form of writing comes to us from these early forms. The Egyptians also invented writing using pictographs or hieroglyphs. That form of writing, although resulting in a rich literature, did not develop into the alphabet type writing we use today.

The Sumerians poured considerable wealth into the construction of clay brick temples and the residences of priests and priestesses who attended to the needs of the gods. The giant temple complexes, Ziggurats (see Figure 4), were centers of economic as well as religious activity. Farmers would bring their produce to the priests and priestesses at the temple who would use enough of the produce to care for the gods and then redistribute it to the people of the city.

FIGURE 4 Schematic of a ziggurat

The great temples were centers of scholarly activity as well. The priestesses and priests controlled the vast set of astronomical observatories spread across the land to observe the stars and planets (mainly for calendar keeping). Our modern day astronomy can trace its roots back to ancient Sumer.

For example, the number sixty, sacred to An (a sky-god), was their basic unit of calculation. Sixty minutes to the hour and 360 degrees in a circle were Sumerian concepts.[23] With each advance in astronomical knowledge came agricultural advances. The priestesses and priests would tell the people when to plant crops, predict lunar eclipses, announce each New Moon (for calendar keeping), and make sundials. The calendar they developed is still used in a modernized form to date certain religious events like Passover, Easter, and Ramadan. All this activity depended upon the systematic astronomical observations made in the network of observatories around Sumer. The highly developed agricultural system and the refined irrigation and water-control systems that enabled Sumer to achieve surplus production also led to the growth of large cities.

The high status of the chief (or en-) priestess was clear. She dominated the religious, scholarly, and commercial worlds, all of which came together in the temple complex that defined the city. In Babylonia, Sumer, and Akkad and their cities the en-priestess was the center figure of the great commercial and scholarly expanses that stretched through the area. The kingly authority was sanctified by her presence.

En'Hedu'anna

Another important Sumerian legacy is its literature. Initially the Sumerians used writing primarily as a form of record keeping. The most common cuneiform tablets record transactions of daily life: tallies of cattle, sheep, and goats kept by herdsmen for their owners, production figures, lists of taxes, accounts, and contracts. But there are tablets of letters and poetry as well. Each letter came encased in a slightly larger baked clay closed container, just as we use envelopes today. Another category of cuneiform writing includes a large number of basic texts which were used to teach future generations of scribes. By 2500 BCE there were schools built just for this purpose with female as well as male scribes[24]. Cuneiform tablets are not large; they are typically less than 25 centimeters on a side. What we would deem 'literature' developed from these early letters and poems.

18

The most famous Sumerian epic, and the one that has survived in the most nearly complete form, is the epic of Gilgamesh. The story of Gilgamesh, who actually was king of the city-state of Uruk in approximately 2700 BCE, is a moving story of the ruler's deep sorrow at the death of his friend and of his consequent search for immortality. We do not know who wrote that great epic. The first poems whose author we *do* know are the wonderful poems of **En'Hedu'anna** (c. 2300 BCE), the en-priestess of the city of Ur. Three long poems to the goddess Inanna, three poems to the god Nanna, and forty-two temple hymns[25] are still found in translation today. She was the only daughter of Sargon of Akkad (2334 – 2290 BCE) who established her in this leading position of the en-priestess. There are now excellent web sites describing her life and works.

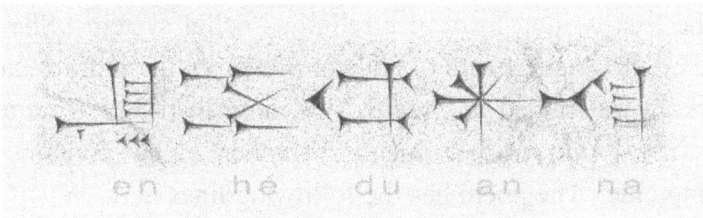

FIGURE 5 En'Hedu'anna in cuneiform

Sargon was the world's first empire-builder, sending his troops as far as Egypt and Ethiopia. He established a unified empire of Sumer and Akkad and tried to end the hostilities among the city-states. Sargon's rule introduced a new level of political organization that was characterized by an even more clear-cut separation between religious authority and secular authority. To ensure his supremacy, Sargon created the first conscripted army, a development related to the need to mobilize large numbers of laborers for irrigation and flood-control works. Akkadian strength was boosted by the invention of the composite bow, a new weapon made of strips of wood and horn.

With our first name, En'Hedu'anna, the tradition of women in science and technology begins. "Hedu'anna" means 'ornament of heaven', the name given to her when she was installed as en-priestess. Figure 5 shows her name in Sumerian cuneiform. We do not know her birth name. She was the chief astronomer-priestess and as such managed the great temple complex of her city of Ur. She controlled the extensive agricultural enterprise surrounding the temple as well those activities scheduled around the liturgical year. Although we do not have technical works from her we know that she was a learned, diversely talented woman of power. And we have her poems. She used her creative talents in the written word, spreading her ideas and beliefs. Her poetry forms the first written form of a religious belief system. She has been called the Shakespeare of ancient Sumerian literature because her works were studied and recited for more than 500 years after her death.[26] One of her hymns, number eight, contains an interesting clue. The poem has the following lines in it:

> in the gipar[27] the priestesses' rooms
> > that princely shrine of cosmic order
> they track the passage of the moon.[28]

There must have been some sort of calendar keeping (astronomy) intrinsic to her position. Another of her poems describes her work

> The true woman who possesses exceeding wisdom,
> She consults a tablet of lapis lazuli
> She gives advice to all lands...
> She measures off the heavens,
> She places the measuring-cords on the earth.

This is the work of a scientist.

There exists an alabaster disk (in the University Museum in Philadelphia) that shows her in a procession. She appears in full religious regalia, the third person from the right on this restored alabaster disc 25.6 cm in diameter (see Figure 6). En'Hedu'anna is our first woman of power and scholarship whose name we know, and the last in a long line of unknown powerful women of the past who followed the stars and the cycles of the Moon. For the next 500 years a daughter of the king was en-priestess of Ur.

FIGURE 6 Alabaster disk showing En'Hedu'anna
Courtesy University Museum

It would be easy to say that En'Hedu'anna was unique. But she was not. There were many such en-priestesses, each a powerful woman who controlled commerce and study. Legend claims that Queen **Semiramis** is the inventor of canals and bridges over rivers and the first to build a tunnel under a river – the

Euphrates – to found the city of Babylon. The legend is probably based on **Sammeramet** who acted as regent of Assyria from 810 – 805 BCE. There are also poets from this part of the world – Inibsari (c. 1790 – 1745 BCE) and Eristi-Aya (c. 1790 – 1745 BCE) who lived in Akkadia. They were daughters of the king of Mari (in Syria), Zimri-Lin. He appointed his daughter, Kiru, mayor of a nearby town.

Around the World

About the same time across the world, on the eastern side of Asia, legend tells us that the first empress of China – **Si-Ling Chi** (c. 2640 BCE) (Lei-Zu) discovered the secret of silk weaving by watching silkworms at work in her garden. She discovered how to unwind the silk from the cocoon, weave it into a garment, and thus ultimately establish the silk industry in China. This was no minor discovery; silk is one of the strongest natural fibers. **Yao**, wife of the fourth emperor, invented spinning. The Empress **Shi-Dun** (c. 105 CE) with her eunuch Cai-Lun was the first to invent a method for making paper from mulberry tree bark. Many women of China were poets (this means they were literate and numerate). **Pan Chao** (50 – 112 CE) (Ban Zhao) was an imperial scholar. She was an official court historian, administered the imperial library, and taught. One of her most famous works is "lessons for a woman". Unfortunately there is a dearth of information about the names of famous women, technical or otherwise, from the past in China. The Chinese were great inventors and had many technical advantages available to them long before they were known in the Europe. For example, the iron plow (6th century BCE) and efficient horse harnesses (4th century BCE) were known in China long before they came to the Europe. Yet, the names of technicians were seldom recorded, and women occupied the bottom of the social and economic structure. The Chinese did record the names of poets and astronomers[29] (all male).

Wandering to India we find **Gargi,** daughter of Vachaknavi, (1500 BCE), honored as a philosopher in the ancient Sanskrit literature, the Upanishads[30]. *Upanishad* means the inner or mystic teaching. They were written by the sages of India between the 8th and 4th centuries BCE. **Maritrayee** was similarly honored in later Hindu writings. Another learned lady was **Khana** of India, assumed to live between 1200 – 800 BCE. Her history is mostly legend, but it is said that her knowledge of astronomy was better than her astronomer husband's.

Returning again to Babylon, **Tappeti-Belatikallim** (c. 1200 BCE) was known as an alchemist who worked with perfume production. She is known as the world's first chemist.

Tradition, as deduced from tomb paintings (7th – 5th centuries BCE), indicates that Etruscan (in Italy) women enjoyed a rare autonomy. It is fascinating to speculate what we might learn if we had documents from that period. In fact the tomb paintings show that the mirrors used by the women had inscriptions on them, thus implying that the women were literate.[31]

Back to Egypt

We started the story in Alexandria, Egypt. There were women of power and leadership in Egypt even earlier than En'Hedu'anna in Sumer. Around 3,000 BCE there was an Egyptian queen **Meryet-nit** who ruled during the First Dynasty of the united Upper and Lower Egypt. Many women influenced the pharaoh even to the point of assuming leadership roles. For example, **Hetepheres II** (c. 2510 BCE) became *Controller of the Affairs of the Kiltwearers*, which meant she ran the civil service, as well as overseers, governors, and judges. Around 1878 BCE

Aganice of ancient Egypt ruled as consort (daughter or sister) of the Pharaoh Sesotris and supposedly was able to predict the planetary positions. The Egyptian queen **Hatshepsut** (c. 1480 BCE – fifth pharaoh of the 18th Dynasty) was also known as a physician. There was **Berenice** who governed Egypt (246 – 241 BCE) when her husband was away. Her court astronomer, Colon, named one of the constellations after a lock of her hair. That constellation is still known as *Coma Berenices* (the hair of Berenice). One of her descendents was **Cleopatra VII** (69 – 30 BCE) today known for her beauty, but in her time known for her diverse talents and intelligence. No mere sex object and fluent in nine languages, she was the last person to rule Egypt as a Pharaoh. She was the first and only Ptolemy who learned the language of her country, Egyptian. It was Cleopatra who introduced the famous Alexandrian astronomer Sosigenes to Julius Caesar, and it was Sosigenes who reformed the Roman calendar which was to last for 1500 years until it was again reformed by Pope Gregory. A Roman Consul described her thusly:

"For she was a woman of surpassing beauty, and at that time, when she was in the prime of her youth, she was most striking; she also possessed a most charming voice and knowledge of how to make herself agreeable to every one. Being brilliant to look upon and to listen to, with the power to subjugate every one..." [32]

And let us not forget **Mary the Jewess**, an alchemist in the first century CE also in Alexandria. She discovered the formula for hydrochloric acid and invented many tools for handling chemicals, one of which is still known today as the *ban marie* – the water bath of Mary (the double boiler), *Marianbad* in German. It is also the prototype for the modern autoclave. She invented a still called the tribikos, which may have been the first device for distillation. Once, while experimenting with sulfur vapor, she synthesized a metal alloy coated with black sulfide, a compound still known as 'Mary's Black'. The *Axiom of Maria* (a precept in alchemy) has been attributed to her: "*One becomes two, two becomes three, and out of the third comes the one as the fourth.*" Carl Jung used this axiom as a metaphor for the process of individuation.

There was also **Beruryah** of the 2nd century CE cited in the Talmud for her scholarship.

These are not many names, but there are enough to excite us on the search. There are very few technical male names from these periods.

The Healers

There is one field where women have always participated — health care. Much of what we call medicine and midwifery is and always has been the province of women. Midwifery was almost exclusively run by women until the 18th century when men usurped the lead away from this traditional women's task. Unfortunately, history rarely recorded their names. Perhaps that is because women have always been physicians so they were too numerous to name.

One of the earliest written names of a woman (or a man) who was a physician is **Merit Ptah**[33] (c. 2700 BCE), a name from 4800 years ago! Her image is on a tomb in the Valley of Kings in Saqqara, Egypt (see Figure 7). She was described by her son, the high priest, as "the chief physician".

The participation of women in surgery began before that – over 5,000 years ago – when surgical instruments of flint and bronze were placed in the grave of Queen **Shubad** of Ur, ostensibly so that she might practice surgery in the afterlife[34].

FIGURE 7 Merit Ptah

Lost in myth is **Agande** (12th century BCE) who Homer tells us was knowledgeable in the medicinal value of plants. The Greek **Agnodice** (4th century BCE) was brought to trial for acting as a physician/obstretrician. The result of her trial was that the medical profession was legalized for all free-born women of Athens. There was a nurse much honored by ancient Rome for her skill in healing and gynecology: **Acca Laurentia** (634 BCE). And six hundred years later women were still known for their healing skills. Ancient Rome had her own physicians – women like **Victoria** (1st century) and **Leoparda** (late 4th century). **Fabiola** (died 399 CE) –a Christian follower of St. Jerome[35] – also practiced medicine. She opened the first hospital for the poor in Rome. The word *medica* was in common use as a word for a physician who was female.

Artemisia II (died 350 BCE), queen of Caria (the southwest of what is now called Turkey), was famous as a botanist and medical researcher. There were several physicians and midwives from the first century BCE Greece: **Sotira** was a Greek physician; **Salpe** was a well-known Greek midwife as was **Olympias** of Thebes and **Metrodora**. **Laïs** was yet another physician in Greece. One woman, **Panthia**, received a tribute from her husband, Glycon (2nd century CE). He was a physician and honored his wife as "not behind me in skill".[36]

Other Sciences

Women have always been healers. What about the other sciences? Two sciences stayed intact as far back as one wishes to go — astronomy and mathematics. They represent the mainstream of pure science, and they, therefore, provide an especially rich source of names. Calendar keeping (crucial to agriculture, religion, and many other endeavors) was the gift of astronomy/mathematics. Before humanity invented writing, we find astronomical based calendar stones and engravings. There are carvings, pictographs, and bones for clues.[37] Since astronomy and mathematics were the earliest scholarly arts, names of astronomers and mathematicians are easier to find than names from other areas. Historical records tend to record the work of the mathematician/astronomer because of its great practical importance. Astronomy and mathematics marched together through the centuries, not really breaking apart until the end of the 19th century.

Other sciences come from differing sources. For example, in earlier centuries chemists were called alchemists. In fact, the names of women appear in a wonderfully diverse set of places. Women are botanists, engineers, physicians, chemists, mathematicians, inventors, explorers, astronomers, agronomists, biologists, physicists, anthropologists, architects, archeologists – a grand list of scientific disciplines ... as well as poets, artists, musicians, writers, singers, mothers, lawyers, activists, laborers, farmers, leaders, fighters. It is much easier to find information about the women in the second part. History more easily records the warrior, the politician, and the poet.

The Women of Ancient Greece

By Homer's time (7th century BCE) Greek women perhaps held a disadvantaged position but nonetheless were capable of ruling in their husbands' absences. They were not considered inferior or incompetent. Homer and other poets tell stories of these strong and active women. In addition, although the Amazons are lost in legend and unproven to exist, many wrote of them. In fact Homer tells us in the *Iliad* that the Amazonian queen **Penthesilea** fought in the Trojan War (c. 1200 BCE) and was killed by Achilles. Legend records that she invented the battle-axe.

Times changed though. In general Athenian women were expected to keep silent, stay at home, and become proficient in the needle and the loom. Aristotle (384 – 322 BCE, the great philosopher whose writings directed European thought for a thousand years) did not believe that women were educable. Sparta, on the other hand, supported the education and development of women. History tells us the final outcome of these two great states. Sparta was eventually defeated by Athens. The status of women in Greece dropped significantly. Their legal rights eroded away to a shadow of what they had been. There are occasional records of women as poets, leaders, and warriors, just as there are for men, but few of philosophy.

Nonetheless, despite the beliefs of Aristotle, some women managed to obtain access to scholars and scholarly pursuits. A very few valued names do exist. Perhaps surprisingly, the ancient Greek philosopher Plato (427 – 347 BCE) wrote about women in his great work *The Republic*[38]. In the perfect state, he wrote, women as well as men needed education – the same education. Plato taught two women in his school: **Lasthenia** of Matineia and

Axiothea of Philus (c. 350 BCE). The philosopher Socrates (469 – 399 BCE) honored **Diotama** as one of his teachers.

Sappho became so well known as a poet that Plato proclaimed her the tenth Muse. She had many pupils in her island home where she ran a school of poetry and music. But we look for those who were philosophers as well as poets. **Hipparchia**, the wife of Crates (c. 470 BCE – founder of the comedic play), authored a lovely sentiment – *I am much stronger than Atalanta from Menelaus because my wisdom is better than racing over the mountain*[39], an early variant of the epigram 'the pen is mightier than the sword'.

Themista, the wife of Leon and a correspondent of Epicurus (371 – 271 BCE), was known as a philosopher in her own right; Themista was even called a female Solon[40]. **Perictione**, a disciple of Pythagoras[41] (c. 569 – c. 475 BCE), distinguished herself by her writings among which are *Wisdom* and *The Harmony of Women*. **Themistocleia** (6th century BC) was a Delphic priestess, the teacher and mentor of Pythagoras. Legend has it that Pythagoras admired Themistocleia to such an extent that he opened his school to women. Whatever the reason, women did attend his school. After his death, the great school of Pythagoras was run by his daughter and his wife **Theano**. Theano not only worked in the areas of physics, medicine and child psychology, but was an astronomer/mathematician in her own right. Her work on the theorem of the Golden Mean (still in use today) and the corresponding Golden Rectangle are considered to be her most important contributions. Her *Life of Pythagoras* is lost. **Arete of Cyrene** (5th century BCE) taught philosophy in a school in Attica. She was the daughter of Aristippus, the founder of the Cyenaic school of philosophy. Even in the time of Boccaccio[42] (1313 – 1375 CE), 1,000 years later, she was still honored as a veritable prodigy of learning, writer of forty books, teacher of over 110 philosophers. Her epitaph

> The splendor of Greece
> The beauty of Helen
> The virtue of Thirma[43]
> The pen of Aristippos[44]
> The soul of Socrates
> And the tongue of Homer[45]

illustrates the high regard in which she was held.

These are but a few of the women of Greece who won renown for their scholarship. Many other women earned their way through the doorway of the hetaerae – the learned companions of men. Such a position does not exist now. These women were considered highly moral and virtuous, free, but unmarried. There were many such women who were respected scholars. Perhaps the most famous was **Aspasia**, the companion of the political leader Pericles (5th century BCE). Her house became a place of relaxation for many famous scholars and leaders of the day. Tradition has it that she was the teacher of Socrates in philosophy and politics and Pericles in rhetoric. Men brought their wives to her for instruction. Did she write some of the great speeches of Pericles? Maybe yes, maybe no. There is every reason to believe that she influenced Plato's ideas on the equality of women.

Known as the witches of Thessaly (1st through 3rd centuries BCE), women such as **Aglaonice**, were thought to "draw down the Moon" because they knew how to predict lunar eclipses[46]. The word "witch" is an epithet given them by later authors, although it is likely Aglaonice was regarded as a sorceress by her contemporaries for her skill in predicting eclipses. Her boasting gave rise to a Greek proverb applied to braggarts "Yes as the Moon obeys Aglaonice".

One small area of the world covering a mere century or so gives us so many names. The 5th and 6th centuries BCE are sometimes called the Axial Ages because so many influential people lived during this time span: Confucius and Lao Tzu in China, the Buddha in India, Mahavira (founder of Jainism) in India, and Socrates, Plato, and Aristotle in Greece. Five hundred years later Jesus lived in Judea, and then six hundred years after that Mohammed lived. It was a busy time in philosophy, and there are many natural philosophers from then whom we still honor today.

"What I have done here no one has done before", En'Hedu'anna wrote. These women are shining lights from our past. Let me step forward in time to shine that light on others.

AS THE YEARS PASS

BCE

3000	Meryet-nit
2700	Merit Ptah
	Si-Ling Chi
2300	Yao
	En'Hedu'ana
1800	Aganice
1500	Hatshepsut
	Gargi
	Maritrayee
1200	Khana
	Tappeti Belatikallim
	Penthesilea
1100	Agande
800	Sammeramet
600	Acca Laurentia
500	Themistocleia
	Perictione
	Theano
400	Arete of Cyrene
	Aspasia
	Hipparchia
300	Diotama
	Lasthenia
	Axiothea
	Themista
	Agnodice
	Artemisia II

200	Berenice
100	Panthia
	Sotira
	Salpe
	Olympias
	Metrodora
	Laïs
	Aglaonice
	Cleopatra

CE

100	Shi-Dun
	Pan Chao
	Mary the Jewess
	Beruryah
300	Fabiola
400	Hypatia
1981	Dr. Ginous Mahmoudi

[1] Quote from En'Hedu'anna

[2] Gerder Lerner, *Journal of American History*, **69**, 1, 1982, pages 7-20

[3] From Science Its history and development among the world's cultures, by C. A. Ronan, 1982, Hamlyn Publishing Group Ltd.

[4] 'Softer' sciences require less mathematics than the 'harder' ones.

[5] Dr. Summers, president of Harvard College in a speech 14 January, 2005. He has since apologized and resigned.

[6] From now on, the word science is taken to mean science/technology/engineering/mathematics.

[7] Today we call this STEM: Science, Technology, Engineering, and Mathematics.

[8] *The Timetables of Women's History*, Karen Greenspan, 1994, Simon & Schuster

[9] See chapter 4 for an outline of women's rights at the dawn of history.

[10] Anyone who read could also handle mathematics. This was true until the late 20th century when literacy and numeracy were no longer inextricably linked.

[11] *Woman in Science*, H.J. Mozans, 1913, D. Appleton and Company

[12] Biographical Encyclopedia of Science and Technology, I. Asimov, 1982, Doubleday

[13] *Men of the Stars* P. Moore, 1986, Gallery Books, NY, NY

[14] CE = Common Era, BCE = Before the Common Era

[15] pronounced Hip-ah-ti'-a in her own Greek language

[16] A compendium of planetary, Solar, and Lunar positions written by Claudius Ptolemy (c. 87 – 150 CE) who also lived in Alexandria

[17] Astrolabes are used to show how the sky looks at a specific place at a given time.

[18] Asclepigenia of Athens was a younger contemporary of Hypatia. Asclepigenia taught in the neo-Platonic school in Athens, which was headed by her father. Upon the death of her father, Asclepigenia, along with her brother and a colleague, inherited the direction of the Academy

[19] Women in Mathematics, MIT Press

[20] Another early technical name was male - Imhotep - the architect of the first pyramid. The first female was En Hedu'anna (c.2354 BCE).

[21] *Goddesses, Whores, Wives, and Slaves*, Sarah B. Pomeroy. The figure is the one discovered by Sir Arthur Evans and known as the Minoan Snake Goddess.

[22] Also early to develop was the hill country west of the Indus Valley.

[23] They also thought the Sun moved around the sky in 360 days. Thank goodness they got it wrong or we would have circles with 365.25 degrees in them.

[24] Amat-Mamu was a female scribe during the time of Hammurabi.

[25] *Inanna, lady of largest heart*, Betty De Shong Meador, University of Texas Press, Austin, 2000.

[26] Fryner-Kensky *In the Wake of the Goddesses*, NY, The Free Press, 1992

[27] Her sacred private quarters in the temple were called the *gipar*.

[28] translation by Betty De Shong Meador, *Inanna, lady of largest heart*, Betty De Shong Meador, University of Texas Press, Austin, 2000.

[29] Interestingly, the Chinese knew that the Sun had sunspots by the 4th century BCE – long before Galileo saw them through his telescope.

[30] These are holy writ to Hindus.

[31] *Women in the Classical World, Fanthan*, Foley, Boymel, Kampen, Pomeroy, Shapiro, Oxford U. Press, 1994.

[32] Lucius Cassius Dio Cocceianus, a Roman consul, c. 200 CE.

[33] *The Timetables of Women's History*, Karen Greenspan, 1994, Simon & Schuster.

[34] Mead K.C. A History of women in medicine from the earliest times to the beginning of the nineteenth century. Haddam, CT: Haddam Press, 1938.

[35] He translated the Bible into Latin – the *Vulgate.*

[36] *A History of Their Own,* Anderson and Zinsser, Harper & Row, 1988.

[37] Astronomy did not grow out of astrology. The science of astronomy predates the art of astrology by several thousand years.

[38] Book V of Plato's *The Republic*, translated by B. Jowett, Random House, The Modern Library, New York.

[39] *A History of Their Own*, Anderson and Zinsser, Harper & Row, 1988.

[40] Solon was "The Great Law-Giver" to ancient Athens – one of the 7 sages.

[41] He established a school of religion/philosophy. He is best known today for the geometric theorem that bears his name – The Pythagorean Theorem.

[42] Italian man of letters

[43] Penelope, the wife of Odysseus

[44] Philosopher who found the Cyrenaic school

[45] *Women of Science*, J. Mozans, 1913, D. Appleton and Company

[46] Peter Bicknell, *Journal of the British Astronomical Association*, **93**, 1983.

CHAPTER 2

THROUGH THE MIDDLE AGES

"…such beautiful minds…"[1]

We leave behind the early scientists like the honored Hypatia and En'Hedu'anna and travel forward in time to the Roman Empire and beyond. Women persisted and succeeded by following their predecessors' shining examples. In fact, as you will see, the eventual growth of the great convents in the late Dark Ages gave women a relatively straightforward path to scholarship.

A Brief Stop at Rome

In imperial Rome women read and carried copies of Plato's *Republic* because it promoted education for women. In general the status of women in imperial Rome was a bit better than their sisters' status in Athenian Greece. It was not unusual for a young girl of plebian parentage to attend elementary school. Upper class men and women had private tutors. Among the attributes of a good wife was her ability to converse intelligently on philosophy and geometry although the true bluestocking[2] (scholar) was probably rare. And, as happened in Greece, many women were known as worthy poets and orators. All over the Roman Empire women of wealth and influence functioned as benefactors and participants in the public world. **Octavia**, the wife of Marc Antony,[3] wrote a book of medical prescriptions. **Cornelia**, wife of Pompey, studied literature, music, geometry, and philosophy. **Valeria Verecunda** (3-4 c. CE) was the first important doctor in her neighborhood of Rome.[4] In the 4th century CE the young woman **Eustochium** traveled to live with Jerome in the Holy Land and help him translate the Bible into Latin, the future Latin *Vulgate.*[5]

Inevitably the Roman Empire began its slow decline. In 337 CE the Roman Empire split into three parts, one for each of three sons of Roman emperor Constantine: eastern (Constantinople), western (Italy and its environs), and central (the Danube areas) empires. In 475 CE Romulus Augustus became the last of the western Roman emperors. He abdicated a year later, and the Western Roman Empire came to an end. Yet the Ostrogoths (East German tribe) who had conquered Rome were not slothful scholars. The daughter of the first Ostrogothic king of Rome (Theordoric the Great), **Amalsuntha** (498 – 535 CE), could converse in Latin, Greek, and Gothic (see Figure 8). She ruled the Ostrogothic (Roman) empire when her father suddenly died. In 535 CE her cousin usurped the throne and had her killed. Less than two years later the Byzantine emperor Justinian invaded Italy to avenge her death, ending the Ostrogothic rule in Italy. Sadly, he also closed the hundreds of year old school of philosophy in

Athens – the wonderful Academy of Plato founded in 387 BCE. Many of the professors went to Persia and Syria. Nonetheless it was a loss to scholarship.

FIGURE 8 ivory representation of Amalsuntha

Despite the loss of the Academy of Plato and despite the destruction of the Great Library of Alexandria, the wisdom of women and men lit with precious, flickering candles of scholarship the Dark Ages that followed the fall of the Roman Empire. Two major religions tended to claim separate parts of the western world.

Christianity spread throughout Western Europe after 313 CE when Constantine's Edit of Milan proclaimed it legal throughout the Roman Empire. Islam dominated Western Asia after 622 CE (the Hegira of Mohammed). In the Far East, the works of Confucius came to prominence in China. The Buddhists grew more numerous and spread out from India to China and beyond.

A Brief Visit to the Rest of the World

Information about the Middle East, Far East, and Africa during this time is very difficult to find.

FIGURE 9 Tower of the Sun and Moon

Sonduk (c. 634 CE), ruler of Korea, built astronomical observatories. One of her observatories, called the Tower of the Sun and Moon, still stands in the old Silla capital city of Kyongju, South Korea (Figure 9).

40

Chinese women were quite inventive. Around 577 CE the first version of matches were invented by the women of the northern Ch'i province in China when they were under siege and needed to start fires for cooking. The modern match waited another thousand years before its invention in 1830 CE. Moving to the Middle East, during the 8th century the medical school at Baghdad (Iraq) had over 6,000 students of both sexes. During the 13th century over 100 women taught at levels equivalent to a professorship at Derviş monasteries (Turkey) in the Islamic world. Names of two such teachers are **Fatima-bint'Abbas** and **Zeynep**, both legal scholars.

Toward the end of the Dark Ages the Islamic world surpassed the Western world in its access to and respect for technical literature. One can easily find information about male Islamic scholars. During the golden age of Islam (8th – 13th centuries) scholars were protected and honored. In general, though, Muslim women were excluded from scholarly endeavors.

It is extremely difficult to find names from India during this time. I found only two names. One is the daughter of the Indian mathematician Bhasharacharya (1114 – 1185 CE). He did many important things in astronomy and mathematics including resolving a problem with the number zero. He was the first to note that division by zero does not give zero; it results in infinity. He wrote several texts on math, one of which he named after his daughter **Leelavati** ("Beautiful"). The book was used to teach her algebra. She was also an excellent mathematician. In 1816 this book was translated into English. The other name is **Raziya**. Shortly after Leelavati lived, Raziya became the first Muslim lady to rule a Moslem state – she was Sultana of Delhi in 1236 CE and very well educated.

The Growth of the Great Convents

Coming back to the decline of Rome, as the Dark Ages drew a curtain of confusion across Europe, centers of learning, abbeys and convents usually associated with the expanding Christian church, struggled into existence. For several hundred years such abbeys became the refuge of women who wished to follow a separate way filled often with scholarship as well as prayer. For example, in France, Queen **Radegund** (518 – 587 CE) forsook her royal state and founded an abbey for women at Poitiers, France. She was quite interested in medicine and had the quite radical notion of washing the patient clean!

By 800 CE many Christian monasteries, abbeys, and convents existed for the devout woman of the Western world. These places often became great settlements that encouraged scholarship as well as piety. Men and women studied together. The abbey women, along with their male counterparts, copied and recopied the studious works of the day along with the rare literature of the past, maintaining them for Western posterity. Many of the women were illustrators as well as copiers of manuscripts. These illuminations on the parchment manuscripts were things of outstanding beauty (see Figure 10).

By joining one of these great settlements women could be free of the intellectual restrictions put on their sex by the secular world. From the 7th to the 11th centuries abbesses held the same power as their male counterparts in both the secular and the religious realm. The larger religious settlements usually owned the surrounding lands as well as the actual convent buildings. Abbesses had the right to attend ecclesiastical synods and assist in the deliberations of national assemblies. Some even had the right to coin money and raise armies. To put this into 21st century terms, the career opportunities for some women during this time were perhaps greater than they are today – a bold statement, but one worth considering.

illumination adapted from Hildegardis Bingensis. *Liber Divinorvm Opervm*. Ed. A. Derolez and Peter Dronke. Turnout: Brepols, 1996.

FIGURE 10

Often the convents and abbeys would conduct schools to which privileged children, both female and male, came. Schooling in those days meant reading and writing in Latin and sometimes Greek and Hebrew as well as a grounding in mathematics.

Ida (c. 570 CE) of Ireland founded a community of nuns who taught a school for small boys including the future missionary, St. Brendan (also known as Brendan the Voyager). In the seventh century in Ely, England **Etheldreda** founded a monastery. She was followed in leadership by her sister, then her niece, and then her niece's daughter. This suggests a very flexible and complex arrangement. Abbess **Aelfthrith** (c. 694 CE) built Repton in Derbyshire, England into a center for education.[6] Abbess **Hilda** (616 – 680 CE) made Whitby, England, a settlement of men and women, into a center of learning for the whole of England. As chief educator she taught theology, grammar, music, the arts, and medicine. During the 7th and 8th centuries there were over thirty English abbesses who ran monasteries. An English nun, **Leoba** (died 779 CE), because of her learned reputation, was sought out by other church leaders for advice. Around 733 CE, her cousin Boniface asked her to help him in his church building efforts in Germany. She agreed and was appointed abbess at Bishofscheim in Germany. For forty years she taught the young nuns there and was sought by many for her wisdom and knowledge. She was friend and counselor to Hildegard, the wife of Charlemagne.

One woman, Hrotsvit of Gandersheim (c. 930 – c. 990 CE) in Saxony wrote verse, history, and the only dramas composed in Europe from the 4th to the 11th centuries.[7] She was allowed her own court, the right to coin money, and to sit at meetings of the Diet – the ruling body. She is usually known by the Saxon equivalent of her self-styled pen name, **Hroswitha**: *hroth* meaning 'sound' and *swith* meaning 'loud' or 'strong' – she who makes a strong sound.

Women contributed their scholarship despite the social shifting sands beneath their feet. **Herrad** of Landsberg (1125 –

1195 CE) was abbess of the convent of Ste Odile in Hohenbourg and compiler-illuminator of the *Hortus Deliciarum* (*The Garden of Delights*). This large folio consisted of 324 pages and 636 miniature illustrations that depicted biblical scenes and allegorical figures. There are over a thousand texts by different authors on different subjects including poems by Herrad herself – a compendium of medieval learning of the knowledge and history of the world intended for the women in her convent. This encyclopedic work covered biblical, moral, and theological material. Figure 11 shows a plate from this work.

Women's contributions stretched over the several hundred years of the Dark Ages and early Middle Ages (and well beyond, of course). For example, circa 1290 CE **Claranna von Hohenburg**, a Swiss nun, was said to be advanced in scientific knowledge.

One truly shining light stands out from the rest – that of **Hildegard** (1098 – 1179 CE) of Bingen-am-Rhein. One of our rare geniuses, she was a mystic nun who wrote volumes of text, composed music, painted (see Figure 12), and ran her convent. She wrote in her journal speaking of her childhood mentor, Jutta:

"This wonderful woman who had guided me in observing the range of positions of the rising and setting Sun, who had had me mark with a crayon on a wall the time and place where the warming sunlight first appeared in the morning and finally disappeared each and every day of my eleventh year."[8]

This is the action of the true thinker.

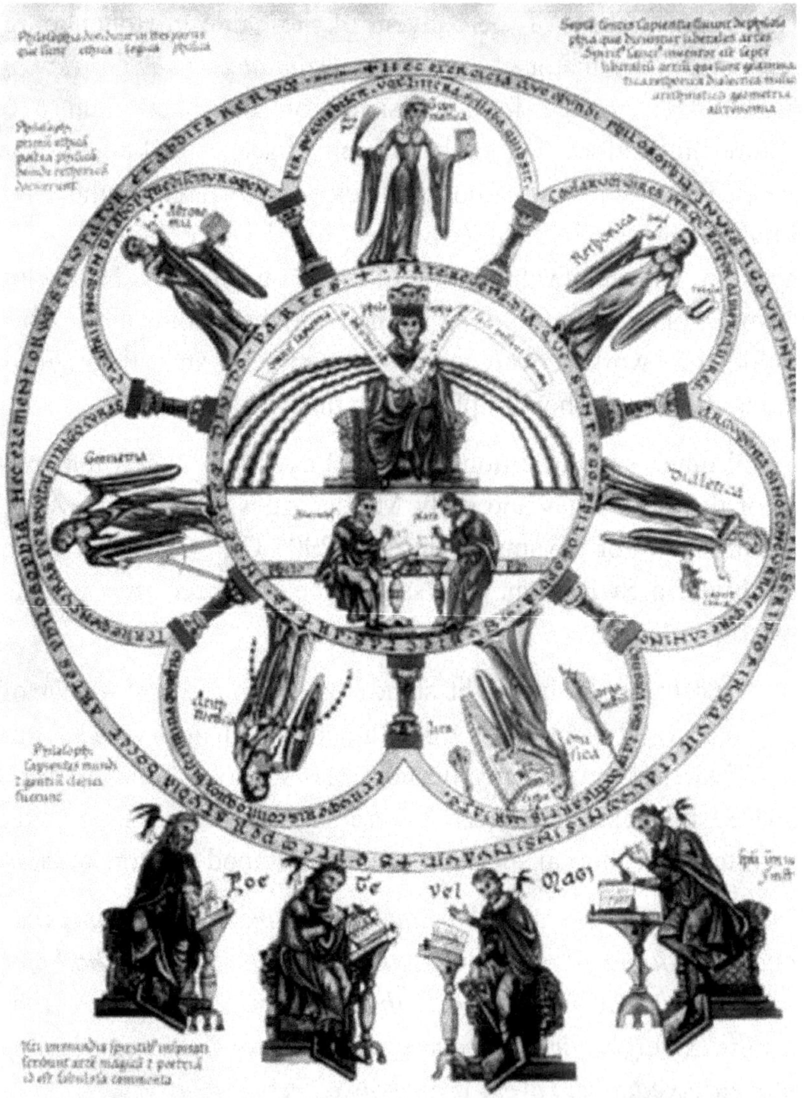

FIGURE 11 The seven liberal arts from the *Hortus Deliciarum*.

46

Her first visionary work was *Scivias* (Know the Ways of the Lord). Literary works followed in profusion: the music and play *Ordo Virtutum* (Play of Virtues – the earliest known liturgical-morality play); *Liber vitae meritorum* (1150 CE) (Book of Life's Merits); *Liber divinorum operum* (1163 CE) (Book of Divine Works). Her *Physica* and *Causae et Curae* (1150 CE), both works on natural history and curative powers of various natural objects, were together known as *Liber Subtilatum* (The book of Subtleties of the Diverse Nature of Things). *Physica* was actually nine books treating minerals, plants, fishes, birds, insects, and quadrupeds. The book on plants has no fewer than two hundred and thirty chapters. Kass-Simon and Farnes have an excellent description of her natural history work in their book *Women of Science Righting the Record*[9].

Her music has been recorded on modern media. She wrote eerily beautiful Gregorian chants (see Figure 12), some so enthralling that it is said people would faint upon hearing her music. She expanded plainchant (a unison chant, originally unaccompanied) a bit beyond the basic intricacies of Gregorian chant even though she had no formal music training.

Her scientific analogies were quite advanced for their time:

"The stars gravitate around the Sun just as the Earth attracts the creatures which inhabit it"

This is the concept of universal gravitation as we know it now, long before it became a standard part of mathematics (five hundred years later by Sir Isaac Newton).[10]

"If it is cold in winter time on the part of the Earth which we inhabit, then the other part must be warm, in order that the temperature of the Earth may always be in equilibrium"

This shows a remarkable sense of the synergy of the Earth. Today we call this point of view thermodynamics. She also believed[11] the Earth was a sphere. She was not alone in this belief; however, the average person thought the Earth was flat.

And foreshadowing Harvey's 17th century theory[12] of blood movement through the body she wrote:

> *"Stars are not immovable but transverse the universe in a manner similar to blood moving through the body."*

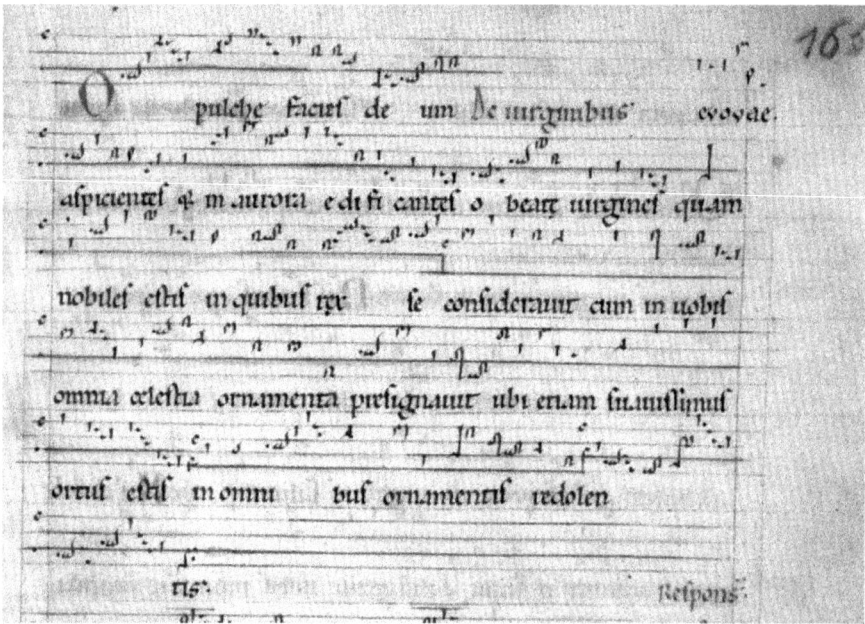

FIGURE 12 adapted from *Hildegard of Bingen, Symphonia Harmonie Celestium Revelationum*, published in modern facsimile by Alamire, POB 45, 3990 Peer, Belgium

She is honored by nurses as the founder of holistic medicine, and delightfully mixed a wonderful common sense with her healing. Here is her recipe for spice cookies (modernized).

"Eat them often," she says, "and they will calm every bitterness of heart and mind – and your hearing and senses will open. Your mind will be joyous, and your senses purified and harmful humours will diminish."

3/4 cup butter or margarine, softened (1 1/2 sticks), 1 cup brown sugar, cream together, add 1 egg. Sift the dry ingredients and add half of them and mix: 1 tsp baking powder, 1/4 tsp salt, 1 1/2 cups flour, 1 tsp ground cinnamon, 1 tsp ground nutmeg, 1/2 tsp ground cloves. Add the other half, mix thoroughly, form walnut sized balls, and bake for 12-15 minutes at 325 degrees.[13]

They are delicious. The recipe is from her book on healing plants in the section on nutmeg.

She believed, as many did, in the healing power of gems.

> "A precious gem will heal the body if taken to bed. And a diamond held in the mouth of a liar or scold would cure any spiritual defects."

Hildegard was a special woman, talented in many, many fields. She corresponded widely with the learned men and women of the day and was highly sought after for advice. She was no shrinking violet in her writings. There is a letter from her to Pope Anastasius IV that begins

> "So it is, O man, that you who sit in the chief seat of the Lord, hold him in contempt when you embrace evil ..."[14]

She was one of the shining lights of the early Middle Ages. Her books were instant bestsellers and beautifully illustrated (see Figure 13).

FIGURE 13 Another of Hildegard's beautiful illuminations

The Healing Arts in the Dark and Middle Ages

As always, women kept their medical tradition alive even in those Dark Ages. For example, in Constantinople during the reign of Emperor Aracadius (400 CE) medicine (and philosophy) gave us the physician **Nicerata**. Jumping ahead a bit – seven hundred years later, interestingly, one of the best equipped hospitals of the time was built in Constantinople by Emperor John II (1118 – 1143 CE). Men and women were housed in separate buildings, each containing ten wards of fifty beds, with one ward reserved for surgical cases and another for long-term patients. The staff was a team of twelve male doctors and one fully qualified female doctor as well as a female surgeon. I don't know their names but they existed. It was not the first large hospital in the area though. In 1096 the first Crusade bought a need for expanded medical facilities in Constantinople. The emperor Alexius built a 10,000 bed hospital/orphanage managed by his daughter **Anna Comena**. She had been well trained by tutors in astronomy, medicine, history, military affairs, history, geography, and math. Running a hospital must have been easy for her. She also wrote a history of her father's life. Called the *Alexiad* this document forms a primary resource for the first Crusade.

Italian women continued to contribute to medicine. The first Western type university was founded in Salerno, Italy in 875 CE as a medical school. And from that time to this, for over a thousand years, women equally with men have been welcome at the doors of Italian universities.[15] So it is not remarkable that there were and are so many talented Italian women. **Jacobina Félicie** (c. 1322 CE) was born in Florence and worked in Paris as a physician. She ultimately lost her battle to practice medicine thus setting a precedent against women attending medical school in France unbroken until the 1800's. No such precedent occurred in

Italy. **Trotula de Salerno** lived in the 11th century (c. 1097) and held a chair in the school of medicine at the University of Salerno. The *Regimen sanitatis salernitatum*[16] contained many contributions from her work and was widely used well into the 16th century. She promoted cleanliness, a balanced diet, exercise, and avoidance of stress – a very modern combination. Her book on the diseases of women[17] was very advanced for the time.

Salerno was home to other women of medicine including **Abella**[18], **Rebecca de Guarna** (fevers, urine, embryology), **Margarita**, and **Mercuriade** (also fevers and embryology) (all 14th century CE). Among those who held diplomas for surgery were **Maria Incarnata**[19] of Naples and **Thomasia de Mattio** of Castro Isiae. **Alessandra Giliana**[20] (1307 – 1326) was an anatomist at the University of Bologna. **Dorotea Bucca** (1360 – 1436 CE) held the chair of medicine at the University of Bologna. **Laura Ceretta** (1469 – 1499 CE) gave public lectures on philosophy. **Battista Malatesta** (1383 – 1450 CE) of Urbino taught philosophy as well. In 1422 **Constanza Calenda** received a degree in medicine from the University of Naples and then taught there. **Calrice di Durisio** (15th century CE) was a surgeon who specialized in diseases of the eye.

Barbara van Roll (1501 – 1581 CE), A Swizz physician, pioneered the treatment of psychosomatic medicine.

Outside the Convents

Fatima of Madrid was the daughter of the astronomer Maslama al-Mayriti (died c. 1007 CE). They lived and worked in Córdoba, Spain. She authored several important works which are known as "Corrections from Fatima". She and her father updated the *Astronomical Tables of al-Khwarizmi*[21] – a mathematician, astronomer, and geographer who worked in Baghdad c. 700 CE.[22]

Italy continued its unique tradition of allowing women into the university. In 1236 CE **Bettina Gozzadini** was appointed to the Chair of Law at the University of Bologna. A century later **Novella d'Andrea** frequently substituted for her father, a professor of law at the same university.

As we move forward in time we find a few women who worked in traditional male occupations. **Isabella Cunio** (13th century CE) may have been the co-inventor with her brother of woodblock engraving. **Fya upper Bach** was a 14th century blacksmith in Siberg, Germany. Twice in her career she held office in the local blacksmiths' guild. **Mary Sidney Herbert** (1561 – 1621 CE), Countess of Pembroke, was quite a learned lady. She was known for her chemistry as well as her poems. One of the leading male chemists of the day – Adrian Gilbert – called her a chemist of note.[23]

I end this chapter with a note about a woman who was not a scientist but was perhaps the first woman in Europe to earn her living through her writing – **Christian de Pisan** (1363 – 1429 CE)[24]. Widowed at twenty-five with a large family to support she took the bold step of going her own way. She wrote at least twelve books and ten works in verse, including what is perhaps the first history of women in the European world – *The Book of the City of Ladies*, published in 1415 CE. This is a fascinating book that combines legend with reality in a sweeping allegory of a city of ladies. Her works are still studied today. I found a tidbit in this book that continues to enchant me. It is a puzzling statement about the education of women.

"God has given them such beautiful minds to apply themselves, if they want to, in any of the fields where glorious and excellent men are active, which are neither more nor less accessible to them as

compared to men if they wished to study them, and they can thereby acquire a lasting name…"[25]

Did she mean that women had the same opportunities as men in the 14th century? I don't know. It is fascinating to speculate. Nonetheless it is always true that most women and men were scrabbling for mere existence in these centuries.

Then Change Sweeps the Lands

The Dark Ages gave way to the Middle Ages as social structures swirled about, settling into new patterns. Feudal Europe became a centuries old war-torn land as rulers continued to fight over boundaries and land rights. At the same time Europeans attempted to retake control of Jerusalem. A series of crusades to Palestine forced open cultural doors long closed. Genghis Khan (1155 – 1227 CE) united the Mongols and swept across the steppes destroying ancient centers of civilization in Eastern Europe. Less than a century later, Marco Polo (1254 – 1354 CE) reached China.

The times leading to the Renaissance were times of extreme even violent change in social customs. The Middle Ages saw the beginning of the substitution of mechanical for muscular power with its accompanying improvement in the human condition. Water wheels, horse collars, windmills, *etc.* all contributed to this improvement.

Until the 14th century men and women of the feudal nobility received approximately the same elementary education. With the rise of the university system in Europe that custom declined. Where religion once supported women in scholarship, it now denied them access.[26] One had to be a member of the clergy to attend a university, and women were denied that right. Several reforms (beginning with those of Charlemagne) were implemented that restricted instruction only to the clergy who were all male.

Charlemagne also decreed that Latin was the official language of learning. Since it was no longer a living language this decree cut women out of learning even more sharply.

Despite Christian de Pisan's statement that women have such beautiful minds, rights for women faded. Let us follow the roller-coaster and travel further along in time.

AS THE YEARS PASS

1st century	Octavia Cornelia
3rd century	Nicerata Valeria Verecunda
4th century	Eustochium
5th century	Amalsuntha
6th century	Radegund Ida Etheldreda
7th century	Sonduk Aelfthrith Hilda
8th century	Leoba
10th century	Hroswitha
11th century	Anna Comena Trotula de Salerno Hildegard Fatima of Madrid
12th century	Leelavati Herrad
13th century	Fatima-bint'Abbas Zeynep Raziya Claranna von Hohenburg

	Bettina Gozzadini
14th century	Jacobina Félicie
	Abella
	Rebecca de Guarna
	Margarita
	Mercuriade
	Maria Incarnata
	Thomasia de Mattio
	Alessandra Giliana
	Dorotea Bucca
	Battista Malatesta
	Novella d'Andrea
	Fya upper Bach
	Christian de Pisan
15th century	Laura Ceretta
	Calrice di Durisio
	Constanza Calenda
16th century	Mary Sidney Herbert
	Barbara van Roll

[1] Quote from Christian de Pisan.

[2] The term originated with the *Blue Stockings Society* – a literary society founded by Elizabeth Montagu in the 1750s. It became a derogatory statement about scholarly women.

[3] This was the Marc Antony who partnered with Cleopatra VII.

[4] *Women in the Classical World*, Fanthan, Foley, Kampen, Pomeroy, Shapiro; Oxford University Press, 1994

[5] a translation of the Bible. For many centuries it was the only translation of the Old Testament (taken directly from the Hebrew) and the New Testament (taken from Greek).

[6] *A history of their own*, Anderson and Zinsser, Harper & Row, 1988

[7] One of her dramas is about a man who sells his soul, predating Faust by 800 years.

[8] *The Journal of Hildegard of Bingen*, B. Lachman, Bell Tower, 1993

[9] Indiana University Press, 1990

[10] Others had also foreshadowed the work of Newton beginning with philosophers in India.

[11] As did a few learned people of the time

[12] Harvey was not the first to discuss it; it was a standard part of ancient Chinese medicine.

[13] http://www.wgbh.org/wgbh/pages/pri/spirit/shows/103recipes.html Bake at 325 degrees. Recipe reconstructed and adapted from Hildegard's c. 1157 treatise *Physica: Liber Simplicis Medicinæ*

[14] *800 Years of Women's Letters*, ed. Olga Kenyon, Penguin Books, 1992.

[15] In 1088 CE Bologna founded what may be called a modern university.

[16] The Salerno Book of Health

[17] Passionibus Mulierum Curandormum (the diseases of women)

[18] Her medical treatises, De atrabile (Black Bile) and De natura seminis humani (Nature of seminal fluid), have not survived.

[19] With respect to the public weal as it relates to the upstanding women of Our [kingdom], We have been attentive and We are mindful in how much modesty recommends honesty of morals. Clearly, Maria Incarnata, of Naples, Our faithful servant, present in Our court has proved that [she] is competent in the principal exercise of surgery, of treatment of wounds and apostemes [tumors]. ... she is found to be competent in treating the above illnesses. *Medieval texts in translation*, K. L. Jansen, J. H. Drell, and F. Andrews, U. Penn Press

[20] She developed a method of draining the blood from a corpse and replacing it with a hardening colored dye, thus allowing the smallest blood vessels to be seen.

[21] From his book on mathematics *Hisab al-jabr w'al mugabalah* we derive the word algebra.

[22] Such tables provided the positions of the Sun, Moon, and planets for future dates – useful for navigation and calendar making.

[23] *Uppity Women of the Middle Ages*, Vicki Léon, Conari Press, 1997

[24] She died in the same year that Joan of Arc assisted Charles VII to be crowned King of France.

[25] *The Book of the City of Ladies*, transl. Earl Jeffrey Richards, Persea Books, 1982

[26] "Astronomy in the monasteries", *New Scientist*, **19**, April, 1984, R. Stephenson

60

CHAPTER 3

THROUGH THE RENNAISANCE AND INTO THE ENLIGHTENMENT

"minds have no sex"[1]

℧he names of women in science and technology from early times are difficult to find; each name is precious. I am sure there are more women to be found, but it is time to move on through the centuries and concentrate mostly on the Western world. This is a conscious decision based on my limited resources which highlight the Western world at the expense of Africa and the Middle and Far East.

The Changing Western World

Lots of things were happening in the West as the High Middle Ages spun to an end. Very briefly, in 1337 CE England's Edward III's claim to the throne of France precipitated the Hundred Years' War (1337 – 1453 CE), a pivotal time that saw the end of 'classic' feudal-based chivalry and the beginning of standing armies and new monarchies. This was complicated by the Black Death – the Plague – that first arrived in Europe in 1348 CE. About one third to one half of the population of Europe eventually succumbed to this devastating illness. The Black Death decimated whole towns bringing cultural change with it because the plague upset the social order. It was a struggle just to survive for the vast majority of people. War-torn lands with their sick and dying populations do not provide much freedom to pursue knowledge. We don't find an excess of Western scholars female or male.

Yet despite the wars and the plague the Renaissance (rebirth) began in the 14th century in Florence, Italy and in the 16th century in northern Europe (*e.g.*, the Elizabethan Age) thus marking the beginning of the long transition from the Middle to the Modern Age. The Renaissance[2] was strengthened by scholars fleeing the struggling Byzantine Empire. In 1440 those scholars founded the Platonic Academy in Florence, Italy, a re-invention of the original Plato's Academy in Athens. In 1453 the Turks conquered the Byzantine Empire, (the fall of Constantinople) causing a further influx of scholars, teachers, and books into Italy. Italian scholars reconnected with the classic past while revitalizing their own culture. Some historians say, however, that the general state of the common folk declined from that of the High Middle Ages (especially perhaps for the women). The Renaissance was not a time of unalloyed good fortune for all.

Johann Gutenberg (c. 1398 – 1468 CE) was part of this rebirth process when he invented printing from movable type. Before Gutenberg developed movable type, Dominican sisters (as well as monks) often did the laborious typesetting for books, and as well, produced the gloriously beautiful illuminations. Gutenberg's work eventually led to a revolution in literacy. By 1480 CE, 110 citites in Europe had printing presses. By 1501 CE, just twenty years later, there were 1000 printing shops in Europe, and an estimated 35,000 titles and 20 million books in print. Accordingly, the literacy rate rose. True paper (invented in China and refined by Islamic scholars) became a useful and affordable commodity. Nonetheless, access to scholarship still remained a privilege accorded to few people.

Medieval guilds were organizations of skilled craftspeople. The term *guild* probably derives from the Anglo-Saxon root *geld* which meant 'to pay, contribute', and thus an association of persons contributing money for some common purpose. Guild members had privileged access to their craft. In cities where guilds were in control the guilds shaped labor, production, and trade; they managed instructional capital and introduced the modern concepts of a lifetime progression of apprentice to craftsman, journeyman, and eventually to master and grandmaster.

The guild of printmaking, a highly skilled trade, attracted women. In 1517, **Caterina De Silvestro** of Naples added italic type to the existing stock of gothic and roman type. Upon the death of her husband, a master printer and bookseller, **Anna Giovanni** of Vincenza not only ran the business but also purchased a paper mill in 1593 (a very shrewd business move). **Charlotte Guillard** (16th century CE) was the first well-known printer who was a woman. She was the widow and wife of two French printers, Rembolt and Chevalon. After the death of her first husband in 1519 she took over management of the print shop and the proofreading of the Latin publications and taught printing to her second husband. Her works were recognized for their beauty and accuracy. Among the publications she printed were a Latin Bible, *Erasmus's Testament*, and *The Works of the Fathers*. Her two volume *Works of St. Gregory* is said to contain only three typographical errors. The scholar Bogard started to write a Greek lexicon. After his death, it was completed and printed by Charlotte. Like the rest of her publications, it was noted for its elegance and accuracy. Her printer's mark was a fancy circle with her initials inside (Figure 14).

FIGURE 14 Charlotte Guillard's printer's mark

The known world got bigger. Just as Marco Polo had re-established contact with the Far East, and the crusades had brought the literature and culture of the Middle East back to Europe, the 15th century had its own travel triumphs. Between 1420 and 1460 Prince Henry the Navigator (Portugal) explored the seas establishing trade routes to India, reconnecting the East and West. Of course, Columbus and Vespucci brought new continents to the permanent attention of Europeans. Magellan's voyagers (1519 – 1522 CE) completed a trip round the world. By the 16th century the entire globe had been criss-crossed with the exception of the poles. People began to travel to strange new places, seeing new species of flora and fauna. This spurred the eventual explosion of interest in natural history and biology (see Chapter 4).

Dutch traders brought Chinese technology to the west. Perhaps one of the most important of these was the breast harness for horses which replaced the throat-girth harness. This vastly improved the agricultural economy. Horses could now pull significantly more weight; more land could be plowed; and transportation of heavy harvests over land became feasible.

Western Christianity was changing as well. Witches (or better, those accused of witchcraft) and the Inquisition played their part in the changes. The horrible witchcraft mania began with the 1487 CE publication of *Malleus Maleficarum* (Hammer of Witches)[3], not to end until the 18th century. Over 80% of the people killed because of this persecution were women. The Inquisition began its insidious march against heretics. These two things would have a devastating effect on women's access to scholarship.

In 1483 CE Martin Luther (founder of Lutheranism) was born. Later John Calvin became a religious leader in Geneva,

Switzerland (1541 – 1564 CE), and the Protestant Reformation was well underway. Slowly, the Roman Catholic Church solidified its beliefs in the face of the growth of Protestant groups, and over time the stultifying religious view of women's inferiority overcame the freeing winds of the convent. The Christian churches (both Roman Catholic and Protestant) consolidated their powers becoming the major claimants in most peoples' lives. Church leaders wrote polemics against women, especially women scholars. Women were considered inherently evil, responsible for the fall of man from grace. The great convents were destroyed or shut down. The profits of centuries of learning and effort were snatched away. Men shut down the convents, destroyed the manuscripts, took the money and lands, and created new male-only universities on the ashes[4]. The Thirty Years War, fought partly for religious reasons, began in Germany in 1618 and in time spread throughout most of Europe. It took especially brave people to stand apart as scholars. Yet, as part of, or despite, these turbulent times, women still managed to contribute to the scholarly life.

Italy Was Special

Something special happened in Italy with the founding of the medical school in Salerno in 875 CE. As I wrote in the last chapter, from that time forward to the modern age, for over a thousand years, women equally with men have been welcome at the doors of Italian universities. I mentioned a few of the talented Italian women in the last chapter, especially those who worked in the medical professions. By the 15th century the Renaissance was in full flower in Italy. **Beatrix Galindo** (1474 – 1534 CE) took her degree in Latin and Philosophy from Salerno and went back to her home in Spain where she became a professor of Latin at the University of Salamanca and filled her idle hours by founding a hospital. We cannot forget **Tarquinia Molza** (1542 – 1617 CE) who excelled in poetry, music, mathematics, and astronomy. She became proficient in Greek, Latin, and Hebrew at an early age. She was so respected that the Senate of Rome conferred upon her the singular honor of Roman citizenship, transmissible in perpetuity to her descendents. During her lifetime she was one of the leading figures of northern Italian musical culture. One can find portraits of her on the web. A good reference for her is a book in Italian titled *Cronistoria del Concerto* by Elio Durante and Anna Maria Martellotti. **Lorenza Strozzi** (1515 – 1591 CE) was born in Florence. She became known for her knowledge of science, poetry, music, and art. As late as 1604 the hymns she wrote were still sung in the churches of France and Italy. At 14 years of age **Fulvia Olympia Morati** (1526 – 1555 CE) wrote dialogues in Greek and Latin. At 16 she was invited to lecture at the University of Ferrara. She died before she could assume the chair of Greek at the University of Heidelberg.

Mozans[5] lists even more Italian women of the 17th and 18th centuries who attained eminence in physical science, mathematics, the classical and oriental languages, philosophy, law, and theology. Seventeen of them are listed below with brief descriptions of each one.

Rosanna Somaglia Landi of Milan was a linguist and translator of the Greek poet Anacreon (who lived between 563 and 478 BCE); **Maria Selvaggia Borghini** (1654 – 1731 CE) of Pisa was a translator of the words of Tertullian (one of the very early Christian writers) and a poet in her own right. There were many scholars like these two who not only understood but also added their own interpretations to the classics.

Elena Cornaro Piscopia (1646 – 1684 CE) of Venice was a prodigy of learning. She received her doctorate in philosophy at

Padua in the presence of thousands of scholars. The University coined a medal in her honor and erected a marble statue of her. Vassar College in New York has a stained glass window depicting her achievements. She studied Latin, Greek, music, theology and mathematics and eventually learned Hebrew, Arabic, Chaldaic (the language of an ancient region of southern Mesopotamia), and also French, English, and Spanish. She studied philosophy and astronomy. Musically talented, by the time she was 17 years old she could sing, compose, and play instruments such as the violin, harp, and harpsichord. She eventually became a lecturer in mathematics at the University of Padua.

Eleonora Barbapiccola (born 1702 CE) of Salerno translated into Italian the works of the French philosopher-mathematician Descartes (1596 – 1650 CE) thus bringing his work to Italy. Her breadth of knowledge in science and mathematics made her famous throughout her region of Italy.

Famous for the phrase *cognito ergo sum* (I think therefore I am) Descartes dedicated his main work, the *Principia Philosophiae* (1644 CE), to **Elizabeth of Bohemia** (1618 – 1680 CE) saying that in her alone were the talents for metaphysics and mathematics united. His work *La Géométrie* includes his application of algebra to geometry from which we now have Cartesian geometry.[6] A number of women were disciples of Descartes including **Anne de La Vigne** (1634 – 1684 CE), **Elizabeth, Princess Palatine** (1652 – 1722 CE),[7] and **Marie Dupré**. Descartes spent his last months in the court of **Christina of Sweden** (1626 – 1689 CE), a patron of the arts and sciences and quite learned herself. All these women were 17th century philosophers. As well thought of as he was, even Descartes moved often to stay ahead of the Inquisition.

Maria Pellegrina Amoretti (1756 – 1786 CE) was a doctor of both canon and civil law, perhaps the first woman to achieve this distinction. **Cristina Roccati** (1732 – 1797 CE) taught physics for 27 years in the Scientific Institute of Rovigo. Her complete lectures on Newtonian physics survive in manuscript today.[8] **Clelia Borromeo** (1684 – 1777 CE) was fondly called *gloria Gennunsium* – the Glory of the Genoese – by her contemporaries because she was so learned in science, math, mechanics, and language. No problem in mathematics seemed beyond her comprehension. The *clelie curve* is named after her (1728 CE).[9] There was also **Diamante Medaglia** (1724 – 1770 CE),[10] a poet/mathematician who wrote about the importance of mathematics in the curriculum of studies for women. Students from all over Europe studied with her. **Anna Morandi Manzolini** (1716 – 1774 CE) held the chair of anatomy at the University of Bologna. She made a number of discoveries as the result of her dissections of cadavers. She made anatomical models out of wax that were highly prized by the University (Figure 15). These models were the archetypes of models used routinely in medical schools today.

During the 17th and 18th centuries Italy was the center of anatomical studies[11]. **Laura Bassi** (1711 – 1778 CE) was an anatomist and natural philosopher who received the doctoral degree from the University of Bologna. She held the chair of anatomy there (appointed in 1732) and also gave lectures in physics (teaching over 28 years). People flocked to hear her lectures. She was an expert in Latin, logic, metaphysics, natural philosophy (Newtonian physics), algebra, geometry, Greek, and French. She and her husband, Giuseppe Veratti, created one of the best experimental physics laboratories known in 18th century Europe. Her scientific papers consisted of one on chemistry, thirteen on physics, eleven on hydraulics, two on mathematics, one on mechanics and one on technology. She did this while bearing eight children. She also introduced Newtonian science to Italy.

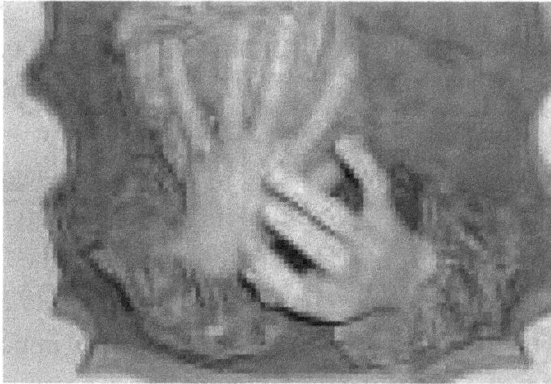

FIGURE 15 Wax model of hands by Manzolini

Archeology in the limited sense is a new science, but the art of collecting rare treasures goes back to the Renaissance and before. As with many other sciences, Italian women led the way. Women like **Elizabetta Gonzaga** (1471 – 1526 CE) Duchess of Urbino, and **Isabella d'Este** (1474 – 1539 CE), Marchioness of Mantua, collected not only antiques in bronze and marble but also rare books and manuscripts in Greek, Latin, and Hebrew.

There were so many talented Italian women who were free to explore their intellectual side. Why was this so? Was this because they followed the example of the ancient Roman matron who insisted upon her rights? I do not know. Perhaps the fact that Italy was the home of the Renaissance is also relevant. Whatever the reason, these strong and learned women of Italy held firm for equality and achieved it. They asked for no favors; they simply expected the same opportunity as a man.

Universities

Salerno was probably not the *very* first university. The Gupta emperors in India founded the University of Nālāndā in the 5th century CE. There were thousands of students and teachers at that school. The courses of study included the scriptures of Buddhism, Vedas, logic, grammar, and medicine. Arguably, however, the first university was the Academy founded in 387 BCE by Plato in the grove of Academos near Athens, where students were taught philosophy, mathematics, and gymnastics. The University of Al Karaouine in Fèz, Morocco is the oldest degree-granting university in the world with its founding in 859 CE by the princess Fatima al-Fihri. Alongside the Qur'an and Fiqh (Islamic jurisprudence) other subjects that were taught were grammar, medicine, mathematics, astronomy, chemistry, history, geography, and music. Zaitouna in Tunis, Tunisia, was another Islamic center of learning. China, too, had its schools. The emperor Wu-Ti (156 – 87 BCE) established a Confucian university for administrators. The first degree-granting universities in Europe were the University of Bologna (1088), the University of Paris (c. 1100) and the University of Oxford (12th century). The last two were restricted to members of the clergy – who could only be male.

The Other Countries

What about the women of other countries? Denied the university freedom accorded Italian women, women of other countries had to have access to private tutors or be self-taught, something only the privileged few could afford. With the invention of movable type printing in the mid 15th century and the introduction of true paper, books slowly became easier to obtain, and with books came increased literacy. In the Protestant churches women were encouraged to learn to read so they could read the scriptures. Despite the exposure, however slight, to literature, many succumbed to the overwhelming pressure to behave as good, obedient wives and thus avoid 'excessive' education. The widespread witchcraft mania that gathered strength through the 16th century kept many women (and men too) from seeking scholarly endeavors.[12] For example, Galileo Galilei (1564 – 1642 CE) ran into his famous trouble with the Roman Catholic Church partly for his insistence that the Earth moved around the Sun.[13]

Nonetheless women continued to argue for the right to study. Despite the barriers women of conviction managed to pursue scientific careers. They must have been extraordinarily brave women (and men). Some struggled against prejudice; some against their church; some even against their fellow scholars.

Margaret of Angoulême, (1492 – 1549) queen of Navarre (in Spain), was in constant correspondence with the learned philosophers of her day and did much to further the cause of the literary movement in France. She created a court whose interests were wide-ranging. Her patronage earned her the title "beloved mother of the Renaissance".

France also saw her first 'mining engineer' in the **Baroness de Beausoleil** (died 1642). She was deeply concerned about the mineral resources of France and foresaw how they could contribute to the country's finance. She and her husband discovered over 150 deposits of iron, gold, and silver. Mozans[14] gives the full title of two of her works (announcing the discovery of ore and minerals useful to the ruler for foreign trade):

Véritable Déclaration de la Découverte des Mines et Minières par le Moyen desquelles Sa Majesté et Sujets se peuvent passer des Pays Etrangers, Paris, 1632.

72

La Restitution de Pluton à Mgr. l'Eminent Card. de Richelieu, des Mines et Minières de France, caches jusqu'à present au Ventre di la Terre, par la Moyen desquelles les Finances de so Majesté seront beaucoup plus Grandes que celles de touts les Princes Chrestiens et ses Sujets plu Heureux de tous les Peuples, Paris, 1640.

They used dowsing as their technique for finding ores and were imprisoned for witchcraft. They died before release, victims of the widespread witchcraft mania.

Princess Anne, the sister of Denmark's King Frederick (c. 1546, probably Frederick II) was a scholar and skilled alchemist.

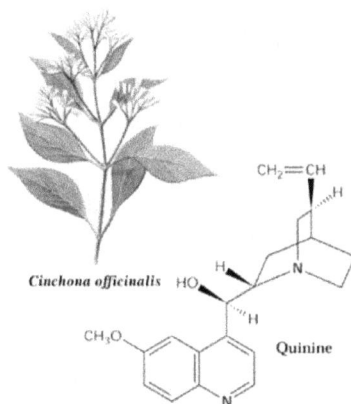

FIGURE 16 Quinine

Ana de Osorio (c. 1630) was the Countess of Chinchon, Spain. Historians are uncertain how quinine got to Europe, but tradition gives the credit to the Countess. While living in Lima, Peru she and her husband predictably came down with malaria.

73

The Countess decided to try a local plant remedy and soon recovered. The plant contained quinine (Figure 16), the miracle anti-malaria drug. She brought the plant back to Spain with her where it quickly proved its worth. The Swedish botanist Carl Linnaeus (1707 – 1778 CE) who developed the scheme for naming plants and animals with fancy Latin names (a classification we still use today) gave the genus name *Chinchona* to several species including the quinine plant in honor of the Countess.

Marie Meurdrac (c. 1666) working in her own private lab in France wrote what is probably the first book on chemistry by a woman for women – *La Chimie Charitable et facile, en faveur des dames*.[15] In it she says that minds have no sex. A facsimile of the cover page of her book can be found on the web.[16]

Minds have no sex! Just as the results of science have no gender, minds have no sex. She was writing about opportunities when she wrote that minds have no sex. She meant that if as much energy was devoted to women's study as to men's study then women would equal men in scholarship. Given the same opportunities, there is no reason for men and women to differ in the results of their scholarship. Neurological differences aside, minds have no sex.

Especially well known in the annals of science is **Maria Sibylla Merian** (1647 – 1717 CE), a natural historian. She studied the flora and fauna of her native Germany and then sailed to Surinam in equatorial South America to study the plant and animal life there (see Figure 17). She returned home where she compiled all her studies into a volume in folio still sought out for its beauty. Her 1705 *Metamorphosis Insectorum Surinamensium* yielded both great beauty and an important scientific discovery. She was the first to record the life cycle of insects, from egg through larvae and pupa to adult. Bolstered by her work it was just at this time that metamorphosis of insects was accepted as possible (as opposed to spontaneous generation). She was quite well regarded by her contemporaries.[17] Linnaeus, the father of taxonomy, referred to her works. At this time most natural history illustrations were often more fancy than fact. There was no photography, and the drawings that artists made during their travels (or from the specimens others sent back) were the only way to record what newly discovered plants and animals looked like. Merian painted on the spot from direct observation, and although there are errors in her work, they are far more accurate than most. They are also some of the most beautiful natural history illustrations ever produced. Today her prints cost many thousands of dollars. Copies of many of her prints (Figure 17) can be viewed with a simple web search. The interdisciplinary Essen Collegium of Gender Studies of the University of Duisburg-Essen/Germany presents the Maria Sibylla Merian Award for outstanding female scientists for their significant contributions to the sciences – in the following disciplines: natural and engineering science, economics, and medicine. Candidates need a PhD as a prerequisite for application. There are no limitations on experience level. The award is endowed with 7,500 Euro. It is sponsored by the Deutsche Telekom AG. An independent, interdisciplinary jury of experts judges and nominates the award winner.

FIGURE 17 Stamps with drawings by Maria Merian

Anna Maria von Schurman (1607 – 1678 CE) learned a dozen languages (among them French, German, English, Italian, Latin, Greek, Hebrew, Syrian, Armenian, Ethiopian), obtained a law degree from Utrecht University, taught philosophy, astronomy, geography, and theology, painted, and sculpted. She was one of the many women who corresponded with Descartes. She wrote a book on Ethiopian grammar and one on gender-neutral intelligence and medicine – a prodigy – and was thought to be the finest scholar in Europe who was a woman. She insisted on education for women and argued for their right to choose whatever subject they wanted to study. She wrote the essay *Whether the study of letters is suitable for a Christian woman* arguing that the study of the arts and sciences were naturally suited to women. In 1646 it was translated into French and in 1659 into English and contains a wonderful statement:

76

"Whatever fills the human mind with uncommon and honest delight is fitting for a human woman."

The *Anna Maria van Schurman* Centre is the inter-faculty platform for gender studies at Utrecht University.

The Healing Arts

Women continued to contribute in medicine even though they typically did not hold a medical degree. In 1628 William Harvey had published his work on the circulation of blood, something foreshadowed by Hildegard (Chapter 2). **Marie Colinet** (c. 1560 – c. 1640 CE) treated patients throughout Germany and in 1580 was the first to use a magnet to remove a sliver of metal from a patient's eye. Two other medical professionals were **Isabelle Warwicke**, an English surgeon (c. 1572 CE), and **Dorothea Christiana Leporin Erxleben** (1715 – 1762 CE) the first woman to receive a full MD from a German university (University of Halle). Hers was an exceptional case, however, and required the intervention of Frederick the Great to make it happen. She took her exams after the birth of her fourth child. Trained originally by her father, the town's physician, she had been practicing as a physician, but without the MD, until she was accused of witchcraft. She replied: "Fine! Here's my dissertation. Let me defend it at the university. Let me take the exams." Officials debated for a year over whether a woman, so often pregnant, could practice medicine. They finally allowed her to take the exams which she passed with flying colors. It was not until 1901 that another woman received an MD from the University of Halle.

A French contemporary of Anna Manzolini was **Mlle. Bileron**. She also fashioned models of the human body. So impressed was the prince royal of Sweden by her work that he saw

77

when visiting Paris that he offered her a position at the royal University of Sweden.

During the Renaissance both men and women practiced medicine. Before 1700 just about all babies were delivered by midwives. Then it changed. From about the beginning of the 17th century through the late 19th century, women were either excluded or banished to such roles as attendant, assistant, or nurse. Some midwives still managed to practice their craft (despite the French ruling against their being MD's). **Margarita Fuss**[18] (died 1626 CE) was so famous and in demand that she was on call throughout Germany, Denmark, and Holland. **Jane Sharp**[19] (c. 1671 CE) was a well known British midwife who wrote the very popular *The Midwife's Book* (still available from Oxford University Press). Her contemporary **Hester Shaw** made up to £1,000 per delivery, an amazing amount of money even today. **Maria Louise Dugès La Chapelle** (1769 – 1821 CE) was a French midwife who studied in Heidelberg and then returned to France to organize a maternity and children's hospital at Port Royal. The earlier works by Trotula (see Chapter 2) were superceded by the works of **Louyse Bourgeois** (1563 – 1636 CE), midwife to the queen of France. **Elizabeth Cellier** was a British midwife (born c. 1640) who became a militant advocate of the education of midwives. She was implicated in a counter plot to the 'popish-plot' to murder Charles II and reestablish Catholicism as the religion of England. Arrested, she spent time in the notorious Newgate prison. Eventually acquitted, she returned to an active life. In 1687 she presented the king with a plan for a college for midwives. These women, however, tended to be exceptions rather than the rule. Except in Italy the practice of medicine was slowly closed to women not to open again until the 19th century.

Why?

That is a good question to which I do not have the answer. As gynecology became more and more 'scientific' in nature, thanks by the way to the work of women, it was usurped by men. It was argued that women lacked the strength and capacity to function as physicians. Of course, midwives are still the most common birth attendant in the world. Birth is an incredibly laborious process requiring a lot of strength and endurance. Women were strong enough to give birth, but not strong enough to diagnose a fever. Yet if women were edged out of professional medicine they remained in other fields.

Natural Philosophers

Sophia (1630 – 1714 CE), the Electress of Hanover, corresponded with Gottfried Wilhelm Leibniz (co-inventor with Sir Isaac Newton of the calculus). Her daughter, **Sophia Charlotte**, queen of Prussia (1668 – 1705 CE) invited Leibniz to Berlin where he founded the Berlin Academy of Sciences. **Marie le Jars de Gourney** (1564 – 1645 CE) was a self-taught intellectual in France who corresponded with Montaigne (French Renaissance essayist) and ended up editing his essays after he died. She composed two outspoken essays: *Égalité des homes et des Femmes* (1622) (equality of men and women) and *Grief des Dames* (1626) (the misery of women). **Anne Bacon** (1528 – 1610 CE) (mother of Sir Frances Bacon, an English scientist) translated many Latin texts into English.

So these women did indeed stay current with the natural philosophy of their day. They corresponded with the great minds of their time. They did as Hypatia did, although many of them did not teach but participated from the sidelines of science by providing sustenance, understanding, and space for the male scholars of their day.

Bathsua Reginald Makin (1600 – c. 1675 CE) was fluent in at least seven languages: English, Latin, Greek, French, Syriac, Spanish, and German. She taught languages in her father's school. She became a tutor, not a governess, to the young Princess Elizabeth of England. She is perhaps remembered best for her essay: *An essay to revive the ancient education of gentlewomen* (1673), a polemic on the education of women citing why education has served them and their country well. Women were still fighting for their right to an education!

Mathematics and Astronomy

Fundamental changes in the world view came with Nicolai Copernicus (1543 CE) and Johannus Kepler (1609 CE). Both advanced astronomical knowledge; Copernicus by publishing the theory that the Sun is the center of the Solar System, and Kepler for suggesting mathematical laws to predict the motions of the planets about the Sun. The Aristotelian world view was disintegrating more and more rapidly. Although the Copernican Revolution represented a true change in world view, it took a few centuries to embed itself into the common consciousness as a successful system. Kepler bravely demoted the prefect circle and found that the 'imperfect' ellipse best represented a planetary orbit. Kepler's three laws are still used today to predict the motions of the planets. Both ideas are fundamental and crucial parts of modern astronomy. Kepler, as many scholars did, moved often, staying ahead of the path of the Thirty Years War and the Inquisition.

Kepler worked for Tycho Brahe, a famous Danish astronomer, known for his precise observational data.[20] Tycho's sister, **Sofie Brahe** (1556 – 1643 CE), trained by her brother as an astronomer, also became a physician and treated patients who came to her brother's observatory. It is said that when "Denmark remembers her Tycho she should not forget the noble woman, his sister who in spirit was more than blood. That shining star in the Danish sky is indeed a double-star."

We come to **Marie Cunitz** (1610 – 1664 CE) an astronomer. Her father educated her at home where she studied languages, classics, science, and the arts. Then she married a physician and amateur astronomer. Before long she became the primary astronomer in the family. At thirty she published a set of astronomical tables (Figure 18). In them she translated Kepler's rather esoteric Latin writings and simplified his method for calculating the positions of planets by omitting his complicated logarithms. Her clear explanations made it an important book, and it went through many editions. In later editions her husband had to write a preface saying it was all her own work. It was so useful that readers assumed he'd written it for her.

Cuniz's troubles didn't end with her death. The 18th century in Germany was not very hospitable to scholarly women. Astronomers of the so-called Enlightenment period couldn't digest her. Forty years after her death, one complained that "she was so deeply engaged in astronomical speculation that she neglected her household." The woman once called the second Hypatia was demoted to second class status.

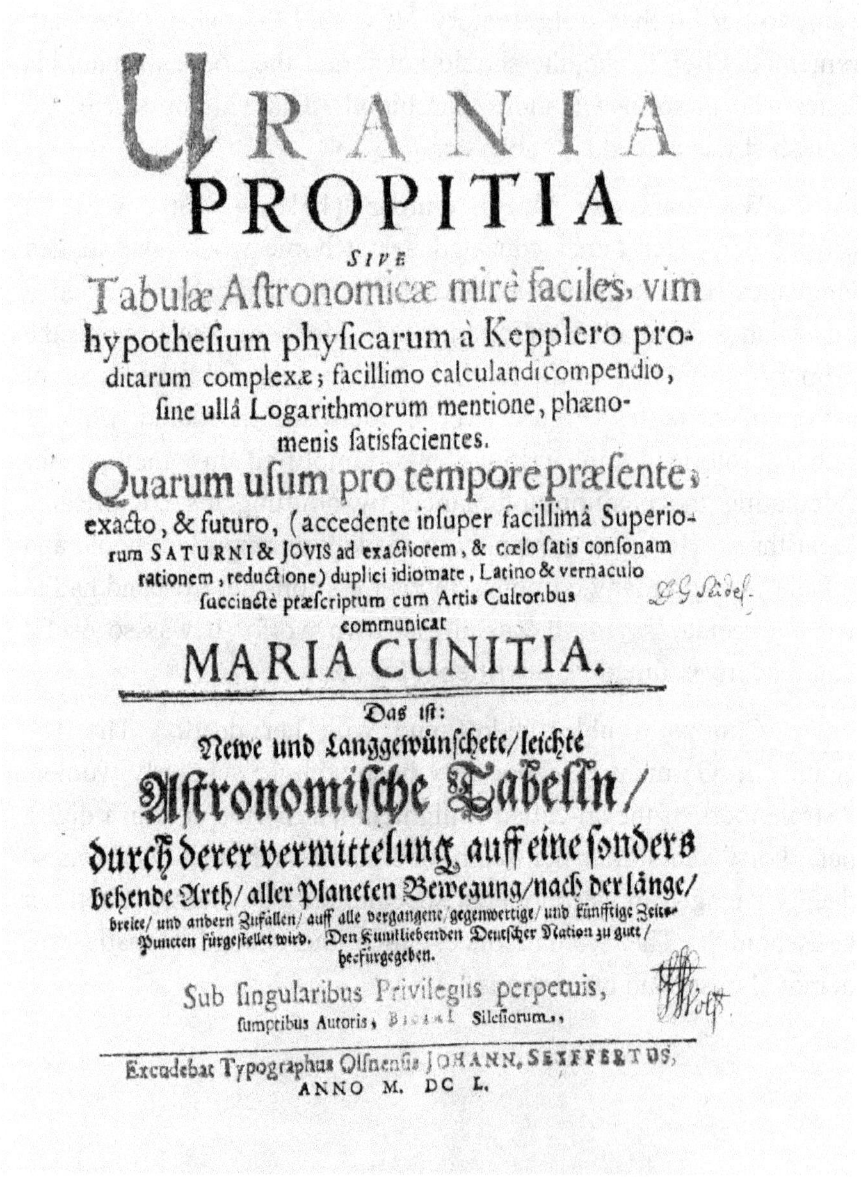

URANIA

PROPITIA

SIVE

Tabulæ Astronomicæ mirè faciles, vim
hypothesium physicarum à Kepplero pro-
ditarum complexæ; facillimo calculandi compendio,
sine ullâ Logarithmorum mentione, phæno-
menis satisfacientes.

Quarum usum pro tempore præsente,
exacto, & futuro, (accedente insuper facillimâ Superio-
rum SATURNI & JOVIS ad exactiorem, & cœlo satis consonam
rationem, reductione) duplici idiomate, Latino & vernaculo
succinctè præscriptum cum Artis Cultoribus *LG Stadel.*
communicat

MARIA CUNITIA.

Das ist:

Newe und Langgewünschete/leichte

Astronomische Tabelln/

durch derer vermittelung auff eine sonders

behende Arth/ aller Planeten Bewegung/nach der länge/
breite/ und andern Zufällen/auff alle vergangene/gegenwertige/ und künfftige Zeit-
Puncten fürgestellet wird. Den Kunstliebenden Deutscher Nation zu gutt/
herfürgegeben.

Sub singularibus Privilegiis perpetuis, *Wolf.*
sumptibus Autoris, Bisant Silesiorum.,

Excudebat Typographus Olsnensis JOHANN, SEYFFERTUS,
ANNO M. DC L.

FIGURE 18 Cover page to the work by Marie Cunitz

Astronomy was the science of choice for other women, including **Maria Margarthe Kirch** (1620 – 1720 CE). She married a Berlin astronomer, Gottfried Kirch. In 1702, as his assistant in observations and calculations she was fortunate enough to discover a comet. Unfortunately custom was not followed in this case; the comet was not named for her.[21] Her husband died in 1710, but she continued to produce astronomical studies including a work on the conjunction of Jupiter and Saturn in the year 1714. Her daughters continued the astronomical work after her death, and they calculated for the Berlin Academy of Sciences its *Almanac* and *Ephemeris*.[22] Such books were valued sources of income for the Academy. Each major city tried to have its own ephemeris. The two sisters of the director of the Bologna Observatory collaborated with him in the preparation of the *Ephemeris* of Bologna. One finds a number of women in astronomical families. In the 1600's in Germany one astronomer in seven was a woman.

Far away in Mexico we find **Sor Juana Inés de la Cruz** (1648 – 1695 CE), a nun in Mexico City, who was a scholar of the court and an astronomer (Figure 19). She is honored today for her poetry and perhaps is the greatest 17th century poet of the American continent. She wrote several secular plays as well. By the age of nine she had mastered Latin. Her personal library had over 4,000 volumes along with many scientific instruments. Dartmouth College has a project underway to present all her works. She believed in the sciences and saw in them no conflict with religion claiming:

> *"It seems to me debilitating for a Catholic not to know everything in this life of the Divine Mysteries that can be learned through natural means."*

When publicly reprimanded by the Bishop of Mexico City for studying and writing, she bravely answered back in a written document that still exists:

"Science and knowledge will strengthen faith in God, not weaken it."

Sor Juana Ines de la Cruz

FIGURE 19 adapted from http://www.edwardsly.com/ines.htm

FIGURE 20 engraving from *Uranographicarum*

Elizabeth Koopman (1647 – 1693 CE), the wife of the Polish astronomer Hevilius (1611 – 1687 CE), collaborated with him on most of his work and who, after his death, edited and published their joint work, the *Prodromus Astronomia*, a catalog of 1,564 stars. Figure 20 shows an engraving from their *Uranographicarum* star atlas. It represents the constellation of Aquarius (with thanks to the Space Telescope Science Institute and the United States Naval Observatory).

Aphra Behn (1640 – 1689 CE) was a British playwright, novelist, and translator of a work on astronomy: *A Discovery of New Worlds* (*Entretiens sur la pluraité des mondes* by de Fontenelle) published in 1686. She was perhaps the first woman in England to earn her living by writing and was honored by burial in Westminster Abby.

Maria Clara Eimmart (1676 – 1707 CE) was expert at precise astronomical drawings. Before the advent of photography drawing was a vital asset to the scholar. Maria was a capable astronomical observer, and between 1693 and 1698, she sketched about 250 scenes of the moon. On May 12, 1706, she observed the total solar eclipse. Some her astronomical paintings ended up in Russia, although some of her drawings are in the possession of the observatory in Bologna, Italy.

Maria Kirch Winkelmann (1670 – 1720 CE) was a German astronomer. She worked with her husband assisting him in his astronomical work. She discovered a comet in 1702. Discovering a comet was an exciting and newsworthy event. They were considered important and unique objects. Computing their orbits was a laborious and difficult task.[23]

By this time the world-changing works of Sir Isaac Newton (1643 – 1727 CE) were making their way through the scholarly world. He codified the law of gravity and showed that it applied to everything on and off the Earth (like the Moon) – the law of universal gravitation foreshadowed by Hildegard (Chapter 2).

Emilie, Marquise du Châtelet (1706 – 1749 CE) was another astronomer-mathematician. She translated Newton's *Principia* into French (c. 1759 CE) thus making his work accessible to her country. It remains the only French translation. Voltaire wrote of her that "two wonders have been performed: one that Newton was able to write this work, the other that a woman could translate and explain it". Her views on the *vis viva* (momentum) opposed those of Newton. Hers proved to be correct. Her own words are prophetic of the burgeoning women's movement.

"Judge me for my own merits, or lack of them, but do not look upon me as a mere appendage to this great general or that great scholar, this star that shines at the court of France or that famed author. I am in my own right a whole person, responsible to myself alone for all that I am, all that I say, all that I do. It may be that there are metaphysicians and philosophers whose learning is greater than mine, although I have not met them. Yet, they are but frail humans, too, and have their faults; so, when I add the sum total of my graces, I confess I am inferior to no one."

Other French astronomers were **Mme du Pierry**, the first women appointed to a professorship at the Paris Observatory (c. 1786 CE), and Mme. **Hortense Lepaute** (1723 – 1788 CE), the wife of the royal clockmaker of France. Mme. Lepaute was hired by Lalande (the Director of the Paris Observatory) to assist him in preparing the orbit of Halley's Comet, due to return in 1759. Such work required prodigious amount of computation, especially to determine the gravitational effect of Jupiter and Saturn on the

Comet's orbit. She also calculated the conditions for the eclipses of 1762 and 1764 for the whole of Europe and published a chart showing the path for every quarter of an hour. She went on to publish other works and continued to work at the Paris Observatory until poor eyesight forced her to stop. She produced the table of the number of oscillations per unit time of pendulums of various lengths for the book that was published under her husband's name, *Traite d'horlogerie*. The beautiful rose *Hortensia* is named for her (Figure 21).

FIGURE 21 *Hortensia*

Mme. du Pierry carried on Lepaute's work and computed tables for the lengths of day and night, and tables of refraction for the latitude of Paris. These tables and catalogs were crucial for the development of modern astronomy. Each one contributed its part to a growing body of work that culminated in such great catalogs as the guide star catalog used to aim the Hubble Space Telescope.

Almost every Earth-orbiting satellite carries aboard a star catalog to orient its way around the sky.

Jeanne Dumée (died 1706 CE) was another French astronomer who tried by her own example to convince men and women that there was no difference in their brains. Her manuscript written in 1680 is still in the National Library of Paris.

She was another woman who believed minds have no sex.

Margaret Cavendish (1623 – 1673 CE) was the Duchess of Newcastle and author of *The Blazing World* in which the heroine makes a round trip of the Moon and planets and thus qualifies as the first fictional female space traveler. She was a colorful figure, nicknamed "Mad Meg", and a prolific and popular author. The diarist Samuel Pepys described her less kindly as "mad, conceited and ridiculous." She published under her own name — a radical and deliberate infringement of contemporary proprieties — a huge body of work encompassing historical treatises, essays, poems, plays, and autobiography. She claimed, "I would rather die in the adventure of noble achievements than live in obscure and sluggish security."

Marguerite de la Sabliere (c. 1640 – 1693 CE) was a friend and patron of La Fontaine, the poet. She received an excellent education in Latin, mathematics, physics, and anatomy from the best scholars of her time, and she typifies the learned French lady of letters and patron of scholars. Like other salonnières of the period, she suffered from ridicule. The literary critic Boileau mocked her as an amateurish pedant who possessed only the veneer of literary and scientific culture. Her telescope was compared to inverted drinking glasses and her Latin phrases allegedly were full of grammatical errors. Several histories of science present her (incorrectly) as one of the first woman astronomers,[24] due to her research undertaken at the Observatory of Paris.

An outstanding scholar of these times was **Maria Gaetana Agnesi (1718 – 1791 CE)** known as "The Oracle of the Seven Tongues" (Figure 22). By the time she was nine she read, wrote, and spoke seven languages. At thirty she produced a text on mathematics "*Le Instituzioni Analitiche*" that earned her the chair of higher mathematics at the University of Bologna, a position she refused preferring to retire to her home and feed the poor. She also produced the solution to a curve[25] heretofore unsolved – still called "the curve of the witch of Agnesi" and found in basic algebra textbooks (Figure 23). They certainly did not think she was a witch. The term comes from a mistranslation of the Italian for "to curve". *Aversiera* is Old Italian for the verb "to curve". This is very close to *avversiera*, a word cognate with adversary (*i.e.* witch).

$$y = a^3/(x^2 + a^2)$$

is the formula she derived, easy for today's computers; extremely difficult in the 1700's.

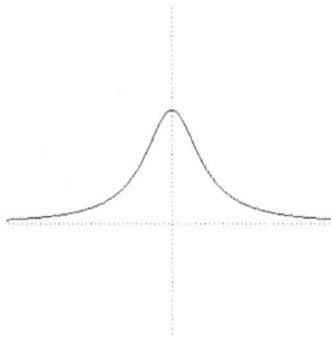

FIGURE 23 Curve of the witch of Agnesi

The French Academy of Sciences would not admit a woman to its august ranks, but M. Motigny, one of the committee appointed by the Academy to report on the work said:[26]

"Permit me, Mademoiselle, to unite my personal homage to the plaudits of the entire Academy. I have the pleasure of making known to my country an extremely useful work which has long been desired, and which has hitherto existed only in outline. I do not know any work of this kind which is clearer, more methodic, or more comprehensive than your Analytical Institutions. There is none in any language which can guide more surely, lead more quickly, and conduct further those who wish to advance in the mathematical sciences. I admire particularly the art with which you bring under uniform methods the divers conclusions scattered among the works of geometers and reached by methods entirely different."

The book made its way into French and English translations. Maria was honored by kings, queens, and popes but steadfastly refused to leave her home in Milan.

Solving a curve may not sound like much. What counts as success in science and technology varies from century to century. We must judge them according to the standards of their own time and place. I cannot omit a woman from this story simply because she would not make the list for the 21st century. For example, to solve a curve as Agnesi did took great skill before we had computers to do the job for us.

FIGURE 22 Maria Agnesi

There *is* something that encompasses not only the 21st century definitions but also all the centuries before it. Successful science works — repeatedly. The results from science can be tested, repeated, and used by others. Successful science works; when the model doesn't work, scientists begin anew to find one that does. Over and over they repeat their attempts until something, even if only the smallest of somethings, works. Small something by small something, the rewards from science accumulate and grow into ever more useful solutions for human problems.

Inventors

Even in the far flung colonies of the Americas women were contributing. They were an inventive lot. In 1715 Sir Thomas Masters of the American Colonies was awarded British patent #401 for an invention by his wife **Sybilla** (died 1720 CE) (he got her name on the patent!) for a method of curing Indian corn. It used hammers instead of gears to grind the corn. He built the device, made money, and became the mayor of Philadelphia. The patent was issued by George I:

> "Letters patent to Thomas Masters, of Pennsylvania, Planter, his Execrs., Amrs. and Assignees, of the sole Vse and Benefit of A new Invention found out by Sybilla, his wife, for cleaning and curing the Indian Corn, growing in the several Colonies of America, within England, Wales, and Town of Berwick upon Tweed, and the Colonies of America."

Eliza Luca Pinckney (1723 – 1788 CE) developed techniques for indigo (an important source of dark blue dye) cultivation in the Carolinas. France had a monopoly on indigo raised in the Caribbean. By developing indigo as a cash crop she created an entire economy for the Carolinas. For more the thirty years South Carolina exported indigo, freeing themselves from British and French control of the product. She began managing her father's plantation at age 17.

Inventors were not limited to the United States. During the time spanning 1631 – 1648 CE one of the most beautiful buildings in the world was built – the Taj Mahal (in Agra, India) – to hold the tomb of Mumtaz Mahal. Her maternal aunt **Nur Mahal** was an inventor. She invented the perfume base called attar of roses – still used in perfumes today. She also invented a method for weaving wool into luxurious cashmere.

As I mentioned before, in general, scholarly pursuits were the peculiar right of the aristocracy. The bourgeoisie did not have the same educational advantage. And with time came the destruction of many of the great convent schools. First they lost their lands and then their independence, and finally the claims to privileges and powers now reserved strictly for men. The nuns were forbidden to teach. At universities outside of Italy the right to study came only with ordination in the Church, a state reserved solely for men. The status of women, outside of Italy, dropped significantly from its level at the height of the $7^{th} - 11^{th}$ centuries.

It seems that women lost rights despite the wonderful shining exceptions. Hopefully we shall see improvements, and Marie Meurdrac's belief that minds have no sex will come to full flower.

AS THE YEARS PASS

15th century	Beatrix Galindo
	Elizabetta Gonzaga
	Isabella d'Este
16th century	Caterina De Silvestro
	Anna Giovanni
	Charlotte Guillard
	Tarquinia Molza
	Lorenza Strozzi
	Fulvia Olympia Morati
	Margaret of Angoulême
	Princess Anne
	Marie Colinet
	Isabelle Warwicke
	Louyse Bourgeois
	Marie le Jars de Gourney
	Anne Bacon
	Sofia Brahe
17th century	Rosanna Somaglia Landi
	Maria Selvaggia Borghini
	Elena Cornaro Piscopia
	Elizabeth of Bohemia
	Anne de La Vigne
	Elizabeth, Princess Palatine
	Marie Dupré
	Christina of Sweden
	Baroness de Beausoleil
	Ana de Osorio
	Marie Meurdrac

	Maria Sibylla Merian
	Anna Maria von Schurman
	Margarita Fuss
	Jane Sharp
	Hester Shaw
	Elizabeth Cellier
	Sophia
	Sophia Charlotte
	Bathsua Reginald Makin
	Marie Cunitz
	Maria Margarthe Kirch
	Sor Juana Inés de la Cruz
	Elizabeth Koopman
	Aphra Behn
	Maria Eimmart
	Maria Kirch Winkelmann
	Jeanne Dumée
	Margaret Cavendish
	Marguerite de la Sabliere
	Sybilla
	Nur Mahal
18th century	Eleonora Barbapiccola
	Maria Pellegrina Amoretti
	Cristina Roccati
	Clelia Borromeo
	Diamante Medaglia
	Anna Morandi Manzolini
	Laura Bassi
	Dorothea Christiana Leporin Erxleben
	Bileron
	Maria Louise Dugès La Chapelle

```
Emilie, Marquise du Châtelet
Mme du Pierry
Hortense Lepaute
Maria Gaetana Agnesi
Eliza Luca Pinckney
```

---

[1] Marie Meurdrac

[2] Today most historians view the Renaissance as largely an intellectual and ideological change, occurring in many places at different times, rather than a substantive one.

[3] The *Malleus Maleficarum* is one of the most famous medieval treatises on witches.

[4] *e.g.* the University of Paris

[5] *Woman of Science*, Mozans, 1913, D. Appleton & Company

[6] The study of geometry using Cartesian coordinates – the familiar x, y, z

[7] Elizabeth married the Duke de Orléans, thus joining the court of Louis XIV of France. Her lively letters are a source of social history.

[8] Newton, of course, is Sir Isaac Newton (1643 – 1727 CE). Cristina was quite current in her teachings.

[9] If the longitude and co-latitude of a point P on a sphere are denoted by θ and ϕ and if P moves so that $\theta = m \phi$, where m is a constant, then the locus of P is a clelie.

[10] http://www.lib.uchicago.edu/efts/IWW/BIOS/A0199.html at the University of Chicago gives a biography of her life.

[11] The first major development in anatomy in Christian Europe since the fall of Rome occurred at Bologna in the 14th to 16th centuries, where a series of scientists dissected cadavers and contributed to the accurate description of organs and the identification of their functions.

[12] Giordano Bruno was burned at the stake in 1600 for his beliefs. He believed that the universe is infinite, that God is the universal world-soul, and that all particular material things are manifestations of the one infinite principle. Bruno is considered a forerunner of modern philosophy because of his influence on the Dutch philosopher Baruch Spinoza and his anticipation of the theories of monism, later advocated by the German philosopher Gottfried Wilhelm Leibniz.

[13] Supposedly Galileo muttered under his breath at his trial, *E pur si muove,* or *and yet it moves* referring to the Earth.

[14] *Woman of Science*, Mozans, 1913, D. Appleton & Company

[15] Chemistry made easy for women

[16] http://chemheritage.org/women_chemistry/know/meurdrac_marcet.html

[17] There is a nice article about her life in *Women on the Margins*, Davis, Harvard College, 1995.

[18] *Uppity Women of the Middle Ages*, Vicki Léon, Conari Press, 1997

[19] *Women in Science*, Ogilvie, MIT Press, 1988

[20] In fact, it was Tycho's data that Kepler used to deduce the orbits of the planets.

[21] Tradition permits the discoverer of a comet to name it, usually after themselves.

[22] An ephemeris is a list of planetary and stellar forecasted positions versus time. Even today this is an extraordinarily complex calculation.

[23] In the 21st century most comets are found by space probes or by amateur astronomers scanning the skies with their telescopes. Discovery is now a bit easier, but the computations for their orbits remain complex.

[24] We know, of course, that we are called astronomers not women astronomers.

[25] To solve a curve is to find the formula for it.

[26] *Woman in Science*, Mozans, 1913, D. Appleton & Company

CHAPTER 4

THROUGH THE 1700's AND INTO THE EARLY 1800's

"turned to the starry heavens"[1]

Women's struggles for education and equality continued through the centuries. Despite the wonderful success stories, it was a bumpy road. Was it a constant grind against discrimination? The previous chapters illustrate an interesting phenomenon. I speculate that women's rights and opportunities for women in scholarship have been on neither a steady incline nor a steady decline. Sometimes up and sometimes down, the opportunities for women varied with time *and* place. Perhaps, just perhaps, the historic past was not always and everywhere an unalloyed dire time for women who wished to be scholars. A brief recap of the opportunities for women might help to make this point.

Pre-history is not the focus of this book; however, although the evidence is slim, prehistoric societies appear to our modern eyes to have been more equalitarian than historic societies[2]. This is partly based on the predominance of female figurines found in archeological sites.

Variations in Women's Rights

The following four maps mark the locations of the ancient places that I mention. At the point when written documents enter the picture (c. 3100 BCE), glimpses filtered through myth and written fragments tell us that women often held positions of authority. For example, it may be that women enjoyed a measure of respect and stature in the sophisticated and technologically advanced urban culture – the Indus Valley Harappan civilization (c. 3000 – 1200 BCE). We cannot be sure of this, however, because the script has not been translated. The Indus Valley was one of the four large ancient civilizations that included Egypt, Mesopotamia, and China.

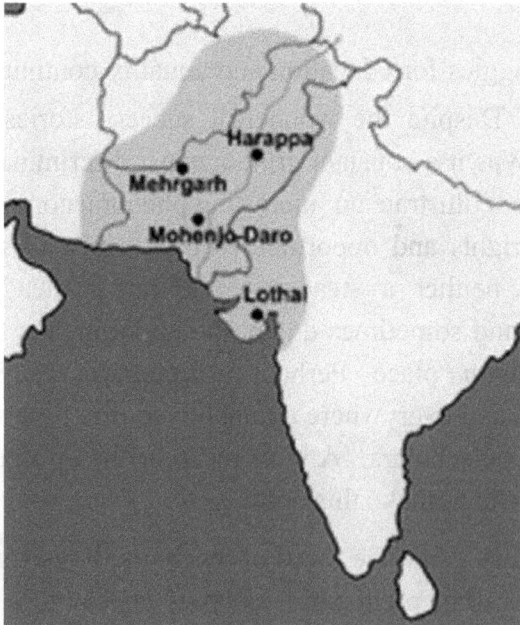

Indus Valley – environs of what is now Pakistan

Likewise it is thought that the Minoan civilization in and around Crete (2600 – 1300 BCE) honored its women. Archaeological evidence suggests that women played an important role in the public life of the Minoan cities. Women were active in trade, sports, and served as priestesses, functionaries, and administrators.

Ancient Egypt treated its women quite well. There are actual documents describing their social, economic, and political lives. Women could manage, own, and sell private property, which included land, portable goods, servants, livestock, and money. They could resolve legal settlements, contract a marriage, and sue for divorce. They could be physicians (*e.g.* Merit Ptah). "Right behavior" (*i.e.* behavior that followed societal demands for stability) was described in guidelines defined by the leaders. For example, c. 2200 BCE "The Precepts of Ptah-Hotep" includes some charming paragraphs[3]:

Be not arrogant because of that which you know; deal with the ignorant as with the learned; for the barriers of art are not closed, no artist being in possession of the perfection to which he should aspire. But good words are more difficult to find than the emerald, for it is by slaves that that is discovered among the rocks of pegmatite...

If you are wise, look after your house; love your wife without alloy. Fill her stomach, clothe her back; these are the cares to be bestowed on her person. Caress her, fulfill her desires during the time of her existence; it is a kindness which does honor to its possessor. Be not brutal; tact will influence her better than violence; her . . . behold to what she aspires, at what she aims, what she regards. It is that which fixes her in your house; if you repel her, it is an abyss. Open your arms for her, respond to her arms; call her, display to her your love.

Similarly the ancient Sumerians (c. 5000 – 2000 BCE) granted women a respected place. Women were free to go to the marketplaces, buy and sell, attend to legal matters, own their own property, borrow and lend, and engage in business for themselves. In Akkad/Sumer aristocratic women might learn to read and write and be given considerable administrative authority. En'He'duanna (Chapter 1) is a perfect example. An Assyrian[4] (post-Akkad – 2nd millennium) marriage contract goes as follows:

Laqipum has married Hatala, daughter of Enishru. In the country [i.e., Central Anatolia] Laqipum may not marry another (woman)—(but) in the City [i.e., Ashur] he may marry a hierodule.[5] If within two years she [i.e., Hatala] does not provide him with offspring, she herself will purchase a slave woman, and later on, after she will have produced a child by him, he may then dispose of her by sale wheresoever he pleases. Should Laqipum choose to divorce her, he must pay (her) five minas of silver – and should Hatala choose to divorce him, she must pay (him) five minas of silver. Witnesses: Masa, Ashurishtikal, Talia, Shupianika.

Infertility was the woman's problem,[6] childlessness was a serious dilemma; children were the desired and expected outcome of a marriage or liaison; however, any subsequent divorce appeared to happen between equals – an interesting *quid pro quo* exchange. Women appeared to have independent access to money.

Moving forward in time to the second millennium, we find that things began to splinter with some areas of the world providing more privileges than others. Where there was equality

102

now there were handicaps. In some places women had rights, in others, none.

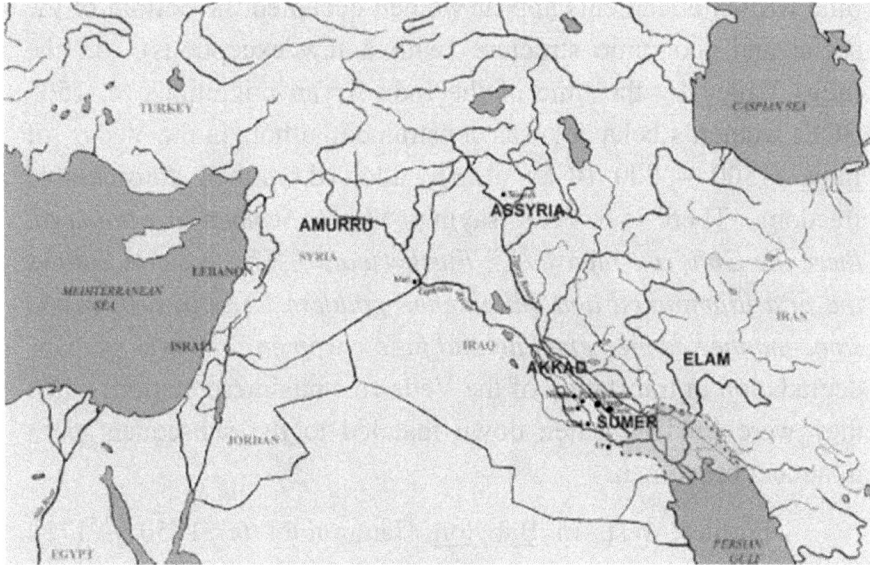

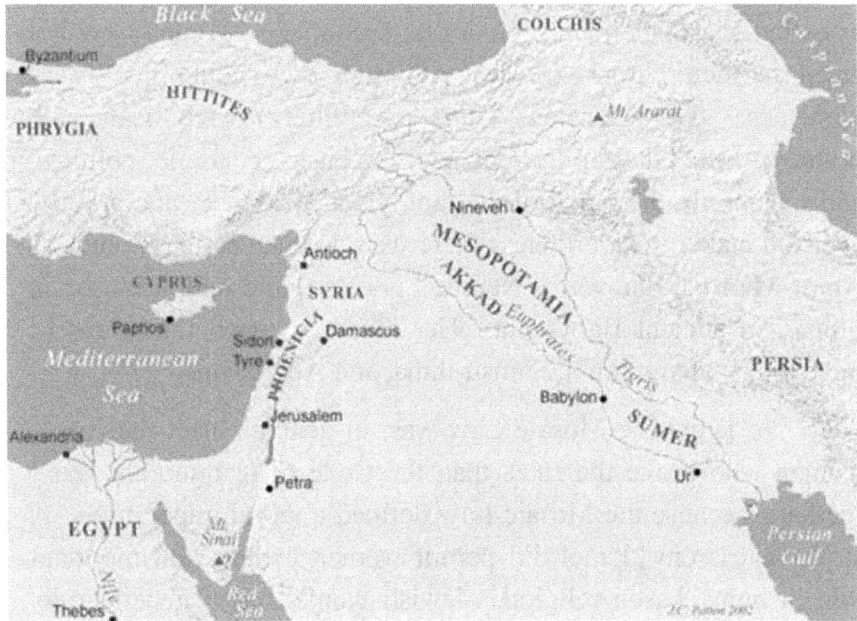

The lack of written documents before the first millennium in China makes it difficult to assess the situation; however, by the time written documents appear women occupied the bottom of the social and economic structure (with a few exceptions). In the Indus Valley by the time of the Indo-Aryan migrations (c. 1500 BCE) women's behavior was constrained; although, the Vedas[7] of India (1500 – 500 BCE) clearly allowed women considerable freedom. There is a Vedic saying: *"Where women are honored, there the Gods are pleased."*[8] Furthermore: *"The sun god follows the first illuminated and enlightened goddess Usha (dawn) in the same manner as men emulate and follow women."*[9] It was perhaps degradation in translation of the Vedas (or misinterpretation) when they were finally written down that led to the subsequent poor treatment of women.

Further west, in Babylon Hammurabi (c. 1750 – 1712 BCE) permitted some protections to women, although not as many as earlier Mesopotamian societies (*e.g.*, Sumer). Women were not equal to men before the law, although they could qualify as witnesses. Transgressions (such as adultery) were dealt with brutally. Since the family remained the basic economic, political, and social unit, the father of the family was the leader and primary decision maker. Nevertheless there were women scribes – such as **Amat-Mamu**. She was a Naditu[10] priestess and temple scribe in Sippar, in ancient Babylonia. Her career spanned the reigns of three kings: Hammurabi, Samsu-iluna, and Abi-eshuh.

In Israel the Mosaic Law was, in general, a bit harsher to women who broke the rules than the Code of Hammurabi was[11] (perhaps because the Mosaic Law defined a moral imperative). A strong patriarchy, Israel did permit women to have an important role in home based religion. Jewish women, as a general rule, were more literate than their compatriots.

Women lost ground in Mesopotamia, but did not do so in the Anatolian plain (or in Egypt). Women were prominent in the Anatolian Hittite state (1600 – 1200 BCE), especially the queen – the Tawananna. Records show that queens conducted treaties and ruled in their own right.[12] Hittite legislation attached great importance to equality between men and women which we learn from written documents concerning property, marriage, and criminal law which applied to the common folk as well as to the nobility.[13,14]

Moving just a bit westward, we encounter the wealthy Linear B culture Mycenae (1600 – 1100 BCE) – covering most of the lands bordering the Aegean, including Troy and the Peloponnesus. Not much is known about this culture although there are a number of active archeological digs. Speaking from a distance of a few hundred years, Homer became a voice for the culture. According to his poems the women were dependent on men; however, they participated in the activities of political, social, and economic life, and could move freely without escort. They felt comfortable offering advice to the men.[15]

In the first millennium the splintering continued. During the centuries that included the Axial Ages (c. 800 – c. 200 BCE)[16] the status of a woman varied widely with the land of her birth; although, in general, women were now considered subordinate to men. Starting again in <u>China</u> we find that women were accorded little dignity and even fewer rights. Female scholars were incredibly rare. Confucius (551 – 479 BCE) codified their low standing; although, both Confucian doctrine and Chinese society at large did acknowledge that women, as mothers and mothers-in-law, deserved respect and with that respect came power within the family structure. Moreover, a handful of extraordinary women managed to acquire literary educations or otherwise achieve positions of far-ranging influence and authority despite social constraints; one became empress – Wu Zetian (690 CE). Chinese technology and invention were quite advanced over what existed in the West; however, women did not participate in its development. Chinese records provide the names of many male scholars and technicians.

In ancient <u>Persia</u> (modern Iran), as opposed to China, women had many rights. We know about Persian women from the clay tablets of the Persepolis Treasury and Fortification records (509 – 438 BCE). The marriage dowry formed part of the marital property owned in common by the husband and wife. Either party could divorce the other, with or without cause, and the dowry returned to the wife. There are records of wives retaining property of their own aside from the dowry, including shares in commercial concerns and real estate, which they sold or traded without reference to their husbands. Upon the death of parents, property passed to the children, with equal shares going to male and female children.

In patriarchal <u>Classical Greece</u> (mainly around Athens) (776 – 404 BCE) women were protected only if they followed the rules which put them completely under male control. Classical Grecian women were constrained far more than their earlier Mycenaean counterparts. Although there were religious outlets for women in Greek society, they had few legal rights. The poet Hesiod (7th century BCE Greece) in his genealogy of the gods tells of the generations of gods first predominately female and then predominately male and goes on to call women "a curse to mortal men".[17] Yet Euripides, thought to be a misogynist, wrote movingly of the plight of women in war in his play *The Trojan Women* (c. 450 BCE). Some women escaped the rigid constrictions; however, escape often meant giving up certain legal protections. Sparta was an exception to this. Spartan women were much freer than their Athenian sisters. Athletic, educated, and outspoken, Spartan women were a respected part of the society.

The position of women improved a bit in <u>Hellenistic Greece</u> (323 – 146 BCE). Hellenistic Greece occupied much of the Aegean Basin and stretched as far as the Alexandrian conquest. [18,19] Women could now leave their homes to buy food and other commodities. They were able to construct marriage contracts that ensured a myriad of legally binding conditions: *e.g.*, the husband shall provide all that is proper for the wife; they choose where they live consulting in common; the husband shall not bring home another wife nor do any evil against his wife. Punishments are spelled out if the terms of the contract are broken.[20] Some women achieved leadership positions and could own and dispose of property. Epicurus (341 – 270 BCE) admitted women to his school as a rule rather than an exception. In some parts of the Grecian Empire women received an education in athletics, music, and reading. Cleopatra VII was an example of the educated and politically savvy Hellenistic woman.

Italian women also enjoyed considerable freedom. Archeological records support the conclusion that Etruscan society (800 – 500 BCE) honored its women. The strait-laced classical Greeks found Etruscan women sexually 'liberated'. As the first millennium came to a close, the women of Republic Rome (509 – 27 BCE) could, in some circumstances, own property, obtain a divorce, and otherwise participate in much of society, although they could not vote or hold office. Education began around six years of age, and in the next six to seven years, boys and girls studied the basics of reading, writing, and arithmetic. However, the head of the family was still the male, the *pater familias*. Cicero described the status of Roman women: '*Our ancestors, in their wisdom, considered that all women, because of their innate weakness, should be under the control of guardians.*'[21] And so every woman had an official (if occasionally unimportant) guardian. During the Roman Empire (27 BCE – 476 CE) women were perhaps more liberated, and they were better off than their Grecian sisters. They had choices. They could influence their society. They demanded rights. In reality, the degree of freedom a woman enjoyed depended largely on her wealth and social status. A few educated women ran literary salons – an activity not considered eccentric. Some women ran their own businesses or had careers as midwives, hairdressers, or doctors.

With the fall of the Roman Empire (5th century) and the end of the *Pax Romana* one might suspect that the status of women would likewise fall. The empire split into two pieces: a western part (Europe) and an eastern part (Byzantium). Soon to join them was the new Islamic Empire.

In the West, during the Dark Ages and early Middle Ages women had access to many of the same privileges that men of equal status had (including scholarship), although, in general, they had to embrace a religious life to accomplish this (see Chapter 2). Aristocratic women exercised considerable power. The laws were flexible and fair to the majority.

In the East, Byzantium was a patriarchal society, although some aristocratic women (Muslim and Christian) exercised economic and political power. Literacy was high, elementary education was widespread (even in the countryside). There was a massive outpouring of books – encyclopedias, lexicons, and anthologies. Many women learned to read and write. A few became physicians. They ran businesses, participated in the early Christian church as nuns or deaconesses, and occasionally took an active role in political affairs. Women had the right to bequeath and inherit property, and married women maintained ultimate ownership over their dowries. In the late Byzantine Era women of scholarship were often referred to as 'second' Hypatias. Her name had become synonymous with a wise and sagacious woman.

In the Middle East, tradition holds that Mohammed (7th century) during his life held women as equals. His sacred texts, which show inconsistent attitudes toward women,[22] were not "sanctified" until a few hundred years after his death. By then Middle Eastern (and Egyptian) women were mostly excluded from scholarly endeavors. Jesus, too, broke with tradition, and apparently treated women as equals. As with the Qur'an, the sanctified Christian writings, with their inconsistent attitudes toward women, came centuries after his death. The result was that the status of women dropped significantly throughout the Middle East.

Ultimately, the Middle Ages evolved into the early Renaissance. Byzantium fell to Islam in 1453 CE marking the end of the remnants of the Roman Empire. Eventually Turkish women were given freedom to work and keep the money they earned, a situation not duplicated in Arabia or the Middle East. Western women lost rights despite the few shining examples of Chapter 3. Witchcraft hunts, misogynistic writings, and the denial of access to education cut sharply into the ability of women to achieve something of worth outside the home.

The family was still the basic unit of society. Any thoughts the average person might have had about 'rights' and privileges was buried by caring for the very old, the very young, and the poor. It is important to remember that the majority of women, like the majority of men, were occupied with the daily struggle for food and shelter. The question of legal rights was unimportant.

This brief description shows that opportunities for women appeared to follow a dizzying roller-coaster ride of human rights and privileges. There were times when and places where women had rights and privileges that far outstripped the inequalities they endured at other times and places. The 'why' of these variations is beyond the scope of this book.

And so we arrive at the 18th century. Happily this roller-coaster began to chug upwards carrying with it the rights of women.

The Sciences Appear

This part of the trip spans the second half of the 18th century and the first half of the 19th century. European and American records are relatively plentiful.

The "Age of Enlightenment" (mostly the 18th century) moved us out of the Renaissance toward the modern age and provided a framework for the American and French revolutions. Immanuel Kant, in exemplifying the age, said *"Have courage to use your own intelligence."* Of course he meant men. However, women as well as men were becoming sure of their own place in the intelligent discourse of humanity. Science was fast growing out of its amateur status to become a studied profession. John Adams (1735 – 1826 CE), second president of the United States once remarked that:

The arts and sciences, in general, during the three or four last centuries, have had a regular course of progressive improvement. The inventions in mechanic arts, the discoveries in natural philosophy, navigation and commerce, and the advancement of civilization and humanity, have occasioned changes in the condition of the world and the human character which would have astonished the most refined nations of antiquity.

The Industrial Revolution spurred by the influx of Chinese technology brought massive changes in agriculture, manufacture, transportation, and technology. The average income increased by ten fold, the population by six fold. Sustained growth became possible.

The application of steam power to printing induced a massive influx of printed works. By 1750 as the literacy rate grew the reading public arrived on the historical scene. The works of Isaac Newton made their way into the general base of knowledge.

111

His fundamental laws of physics were applicable at all levels from the universe at large to the minutest fragment. This universality was a new thing to science (although not to Hildegard). Observation, experience, and experiment became the lodestones of science.

This time is unique because now is when we find the modern sciences peeking through the curtain of history, revealing their distinct patterns. With the onset of the 18th and 19th centuries we find the once complete in itself 'natural philosophy' forming into separate branches of study; *i.e.*, astronomy, mathematics, physics, biology, chemistry, meteorology, geology, and the social sciences, all in various combinations, and all forming the natural and social sciences. This word *science* means the same thing as Hypatia's *natural philosophy*. The separation of natural philosophy into various scientific disciplines occurred relatively recently (in the last few centuries before the twenty-first). Each emerging discipline formed part of the great interwoven tapestry of science held together by the mainstream (astronomy/mathematics). Each strand also brought with it the efforts and particular strengths of the non-specialist. For example, physics had been more a practical skill than a scholar's tool until the 19th century added mathematical rigor to the technical ability of the amateur practitioner. It then grew into the great mix of physics that we have today: *e.g.,* solid state, nuclear, quantum, crystallography, *etc.* Botany, biology, and zoology drew from natural history. Chemistry is almost a direct descendant of alchemy. Similar situations existed for all of the sciences. We lost the generalist and gained the specialist. By this time, instead of drawing from all fields of science and nature (as did the natural philosopher), a scientist was typically a person with a single specialty. These scientists were no longer knowledgeable about the whole field of natural philosophy. There were still some generalists (such as Benjamin Franklin and Jane Marcet), but mostly we shall find that each scientist worked in a specific specialty.

The queen of science, though, remained astronomy-mathematics. We can track these scientists more easily than the others because they were always called astronomer-mathematicians. The others were called natural philosophers, scholars, physicians, midwives, inventors, or alchemists, *etc*. Whereas the field of astronomy-mathematics tracked its own history and recorded the names of its practitioners, the other disciplines were not quite so diligent.

I begin this part of the story with two of the few remaining generalists, **Jane Haldimand Marcet** (1769 – 1858 CE) and **Claudine Pouliet Picardet** (c. 1770 – c. 1820 CE). Jane Marcet was a British writer on natural philosophy. Her book *Conversations on Chemistry* was studied by Michael Faraday, himself a well-known physicist who gave us, among other things, "Faraday's Laws" still used in physics today. Her book went through 16 editions in England and formed part of general chemistry instruction. She suggested that experimentation be part of any lecture on chemistry – perhaps the first to suggest this. Her suggestion is now *de rigor* for all chemistry courses. Her other *Conversations* were on political economy, natural philosophy, and vegetable physiology. The other lady, Claudine Picardet, lived in France and translated several technical works into French. She was typical of the 18th and 19th century woman of letters (generalist) who understood what she read so she could translate difficult and esoteric works and add commentary and corrections of her own.

Princess **Yekaterina Romanovna Vorontsova-Dashkova** (Russian) (1743 – 1810) was the closest female friend of Empress Catherine the Great and a major figure of the Russian Enlightenment. Her name was often spelt in English as Princess Dashkov. After studying mathematics at the University of Moscow she traveled widely throughout Europe (even meeting with Benjamin Franklin in Paris in 1781). Franklin invited Dashkova to become the first woman member of the American Philosophical Society. In 1782, Dashkova returned to the Russian capital, and was at once taken into favor by the empress, who strongly sympathized with her in her literary tastes, and especially in her desire to elevate Russian to a high place among the literary languages of Europe. She was appointed Director of the Imperial Academy of Arts and Sciences (now known as the Russian Academy of Sciences). Dashkova was the first woman in the world to head a national academy of sciences. Although not a scientist herself, Dashkova restored the failing institution to prominence and intellectual respectability. As its director, Dashkova made Benjamin Franklin the first American member of the Imperial Academy of Sciences and Arts in St. Petersburg. They both exemplified the ideals of the Enlightenment that flourished in Europe and America. In 1784 Dashkova was also named the first president of the newly created Russian Academy. In this position, too, she acquitted herself with marked ability. She launched the Russian Academy's project for the creation of its 6-volume Dictionary of the Russian Languages, arranged its plan, and executed a part of the work herself. In 1783 she was elected an honorary member of the Royal Swedish Academy of Sciences, the first woman among this academy's foreign members, and its second female member after Eva Ekeblad.

Founded in 2010 in Edinburgh, the Princess Dashkova Russian Centre builds on a long history of academic contacts and exchange of ideas between Scotland and Russia

Women began to organize. What is thought to be the first scientific society exclusively for women was founded in Middelburg, in the south of Holland in 1785. This was the *Natuurkundig Genootschap der Dames* (Women's Society for Physical Knowledge). Jacoba van den Brande was the first director. It lasted until 1887.

Before I discuss the more easily found astronomer-mathematicians, there are other technical women to share, each bringing her own strand to a newly developing area of science.

Biology and Related Sciences

The budding science of biology was fortunate to have had many women contribute, bringing botany, horticulture, plant physiology, agronomy, zoology, and natural history into the ever widening field of biology. Rather nicely, in 1822, New York botanist/geologist Amos Eaton said "I believe more than half the botanists of New England and New York are ladies." A rigorous classification system of plants and animals had just been defined by the Swedish botanist Carl Linnaeus (1707 – 1778 CE), the father of taxonomy, who arranged plants and animals into natural systems reflecting their common relationships. All that memorizing of class, genus, and species one does in biology class started with Linnaeus. The opening of the world to exploration that occurred in earlier centuries brought with it exposure to strange new flora and fauna, all of which had to be classified.

Botanists like **Jane Colden** (1724 – 1766 CE) were educated at home. Her catalog (classification) work was extremely accurate. She published the first illustrated flora of New York State, and she was the first to classify plants according to the Linnaean system. So Jane was right on the mark with her work, and botany moved to the springboard into modern biology. In 1963 The New York Garden Club of Orange & Duchess Counties, published, on the 50th anniversary of The Garden Club of America, a reproduction of her original manuscript. What began in the drawing rooms of talented artists ended in the great biology laboratories of the 21st century.

There were, as well, horticulturists like **Martha Daniell Logan** (1704 –1779 CE) who wrote a standard text on horticulture – the *Gardener's Kalendar*. It was the first American work on gardening. Even plant physiologists[23] existed – such as **Agnes Ibbetson** (1757 – 1823 CE). *The Journal of Natural Philosophy, Chemistry, and the Arts* published several of her works (1809 – 1813) on the microscopic structure and physiology of plants.[24] Copies of the pages are provided on the web site of Michigan State University. Women seemed quite at home in these types of pursuits. **Martha Laurens Ramsey** (1759 – 1811 CE) was an agronomist (the study of agriculture enabled by science). She experimented with olives as a cash crop in South Carolina.[25] **Margaretta Hare Morris** (1797 – 1867 CE) was a naturalist known for her work on the life cycle of the 17-year cicada and was the first woman to become a member of the Philadelphia Academy of Sciences. **Maria Angela Ardinghelli** (1728 – 1825 CE) of Naples was famed for her profound knowledge of physics and mathematics and produced an Italian version of Stephen Hales' *Vegetable Statics*.[26] This book demonstrated that plant leaves absorb air that is used for the plant's nutrition. She recalculated his results, corrected his mistakes, and redid his experiments.

116

Josephine Kablick (1787 – 1863 CE) of Bohemia studied under the best botanists of her time. She was an indefatigable explorer, as well as an enthusiastic collector, and many public institutions owe their best samples to her endeavors.

Many women worked alongside their husbands and brothers illustrating their work with delicate yet thorough and accurate drawings thus contributing silently and beautifully to the advance of the field. The sciences of botany and zoology are, in part, a natural outgrowth of the painting and drawing (and close attention to detail), skills taught to many women.

FIGURE 24 a mollusk

Jeanne Villepreux-Power (1794 – 1871 CE) was a zoologist. Born in the French village of Juillac, she walked from there to Paris (almost 300 km) where she became a brilliant dressmaker's assistant. Through her fame as a seamstress, she met a rich, noble English merchant in Sicily, James Power. They were married in 1818 in Messina and lived in Sicily for 20 years. The shoemaker's daughter soon became a naturalist, fully self taught, and she was first (1832) to create and use aquariums for experimentation in aquatic environments. She authored *de Observations et experiences physique sur plusieurs animaux marins et terrestres*[27] as well as several other papers. She studied fossil shells and the humble mollusk (Figure 24) as well as Argonaut shells and is perhaps most famous for inventing the aquarium. She invented three kinds: one of glass to study live mollusks (classic aquarium); another also of glass surrounded by a cage submerged in the sea to study small mollusks, and a third kind of cage for larger mollusks which could be sunk and anchored at a given depth in the sea. She carried on her work at Messina in Sicily. She was, from 1832 – 1842, the only woman in the *Catania Accademia* and a corresponding member of the London Zoological Society (plus 16 other academies).

FIGURE 26 16 sided house[28]

Architecture

Architecture is an old subject, of course, and women were part of it starting with the legendary Semiramis (Chapter 1). **Jane and Mary Parminster** (c. 1795 CE) were architects who designed a famous octagonal-centered 16-sided house in near Exmouth in Devon, England. The house now belongs to the National Trust (Figure 26).

Leaping ahead one hundred years, **Harriet Morrison Irwin** (1828 – 1897 CE) was the first woman in the United States to patent an architectural design. She wanted a house that was easier to dust and clean, so to avoid the 90° corners in rectangular houses, in 1869 she received a patent for a hexagonal building. She built the house which still stands today in Charlotte, North Carolina.

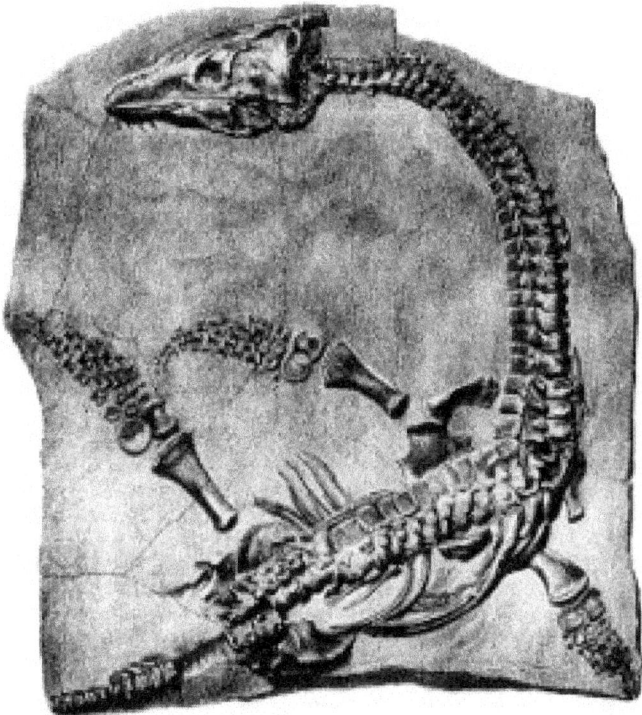

FIGURE 25 Plesiosaurus

Paleontology

Paleontology grew out of a fascination with fossils, and fossil sequences were noted by the early 18th century.[29] **Mary Anning** (1799 – 1847 CE)[30] was an amateur who found the first complete ichthyosaur skeleton in 1811 and the first British pterodactyl as well as the *Squaloraja* fossil fish (a transitional link between sharks and rays) and finally the *Plesiosaurus macrocephalus* (Figure 25). Lady Harriet Silvester[31] wrote:

"The extraordinary thing in this young woman is that she had made herself so thoroughly acquainted with the science that the moment she finds any bones she knows to what tribe they belong. . . . by reading and application she has arrived to that greater degree of knowledge as to be in the habit of writing and talking with professors and other clever men on the subject, and they all acknowledge that she understands more of the science than anyone else in this kingdom.[32]"

Schooling

The chance at official schooling was finally coming back to women. Schools for women in Britain were common, and some included technical subjects. For example, the British schoolmistress **Margaret Bryan**[33] (born c. 1760) insisted on a curriculum that included astronomy, mathematics, and science. A male professor of mathematics commented on her published lecture notes by writing

"I have read over your lectures with great pleasure and the more so, to find that even the learned and more difficult sciences are thus beginning to be successfully cultivated by the extraordinary and elegant talents of the female writers of the present day".[34]

Schools for women in the United States were slower to come, but come they did. In 1654 **Marie Guyart, Marie de l'Incarnation** (1599 – 1672 CE)[35] opened the first convent and school for girls in the American continent, located in what is now known as Quebec City. As an Ursuline nun, she left her native France and her son to establish a convent in the New World. The daughters of French colonists and Amerindian girls were taught there. She learned Algonquin, Montagnais, Huron, and Iroquois and wrote dictionaries for these languages. The school still exists today.

Abigail Adams bemoaned the lack of education for women in a letter to her husband John in 1778: *"You need not be told how much female Education is neglected, nor how fashionable it has been to ridicule Female learning."* Between 1790 and 1830 opportunities improved. Sometime between 1750 and 1850 the literacy rate for women in the United States rose from a grim half that of men to be almost equal in part due to the appreciation of education for both men and women. It was slowly recognized that competence in a woman was a desirable quality, and that required that she reverence herself (including her mind).

However, even though schools for women opened, a few important universities hung back from admitting women. For example, Yale University did not admit women until 1969. Fortunately Yale became the exception rather than the rule. Interestedly, in 1783 **Lucinda Foote** tried to enter Yale. Her mental ability so astonished the president of Yale that he gave her a testimonial:

"Let it be known unto you, that I have tested Miss Lucinda Foote, aged 12, by way of examination, proving that she has made laudable progress in the languages of the learned, viz, Latin and

the Greek .... I testify that were it not for her sex, she would be considered fit to be admitted as a student..."[36]

Eventually schools and colleges for women sprang up around the United States. **Mary Lyon** (1797 – 1849 CE) helped to found Mt. Holyoke College (1836). The Mt. Holyoke College web site has a nice article on her life. She herself taught chemistry at the College and required seven courses in science and mathematics from all students for graduation, something that today is quite rare. Mt. Holyoke College became an important resource for women who wished to pursue a research career in science. She chose South Hadley, Massachusetts because the town offered $8,000.00 in support.

Emma Willard (1787 – 1870 CE) was a scholar who believed as many did before her that women could learn as well as men could. At age 17 she taught herself geometry. In 1821 she established a school in Troy, New York to teach women science and mathematics as well as the arts. The Common Council of Troy voted $4,000.00 to support the school. The school still exists. She wrote some of the first American history and geography textbooks. Anticipating modern education theory she advocated active learning.

The Patapsco Female Institute (Maryland) opened its doors in 1838. The principal **Mrs. Mary Norris** insisted on a curriculum that included English, the classics, languages, natural and abstract sciences, modern history, chemistry, botany, piano, painting, and guitar. In 1841 **Almira Hart Lincoln Phelps** (1793 – 1884 CE), the sister of Emma Willard, took charge of the school which soon became financially successful (see her portrait – Figure 27). She was a botanist who in 1859 became the second woman (following Maria Mitchell – see later in this Chapter) elected to the American

Association for the Advancement of Science. In 1829 she published a textbook, *Familiar Lectures on Botany,* which enjoyed wide use and went through nine editions in 10 years. The online *Encyclopedia Britannica* has a nice article about her.[37] The site of the school is now an active archaeological site and is on the National Register of Historic Places.

The second half of the 19th century saw an explosion in education opportunities for women. The Georgia Female College in Macon, Georgia (now known as Wesleyan) granted the first Bachelor Degrees (including science) to women in 1840. Antioch College in Ohio was the first non-sectarian college to grant women the same rights as men. **Jane Andrews** was the first student to register there in the fall of 1853.

FIGURE 27 Almira Hart Lincoln Phelps
By permission of the Patapsco Institute

Chemistry

Women were, as they had throughout history, discovering their own strengths. Even without access to colleges women joined those people forming the new strands of the sciences. For example, modern chemistry wove itself into the tapestry of modern science in the late 1700's. The wife of Antoine Lavoisier **Marie Anne Pierrette Lavoisier** (1758 – 1836 CE) worked side-by-side with him, so much so that it became difficult to disentangle his work from hers. Married to him at thirteen, she spent her teen years working and learning with him. She even learned English to keep him abreast of chemistry in England. He is sometimes called the "father of modern chemistry",[38] she the "mother of modern chemistry". Following the new trend of measurement of actual quantities in a chemical reaction, they gathered enough data to verify the law of conservation of mass in a chemical reaction. They also showed that water was made of hydrogen and oxygen, naming oxygen from the Greek word, οξύς–γενής, meaning acidifying. Antoine was killed in the French Revolution, and after the times had settled down, Marie married the physicist Sir Benjamin Thompson, Count Rumford, a marriage that lasted only four years. For many years afterward she ran a scientific salon.

A British chemist, **Elizabeth Fulhame** (c. 1794 CE), conducted experiments in combustion and then developed a theoretical explanation for them. That, in itself, was a rare thing – to be both an experimentalist and a theoretician. Even today, people tend to be one or the other, not both. She was elected an honorary member of the Philadelphia Chemical Society, and her book *Essay on Combustion* was reprinted there in 1810. She is credited with the first recorded experiments on photochemical imaging, the first proposal of a two-step chemical reaction, and the first published concept of a catalytic process, and her discoveries were acknowledged by leading chemists in both the United States and Europe.[39] This was important because most Europe scientists considered American science second-rate.

Modern archeology also joined the tapestry. The first woman to take part in an archeological dig was the youngest sister of Napoleon Bonaparte, **Caroline** (1782 – 1839 CE). She spent a great deal of time at the excavation at Pompeii (c. 1808 CE). She did much to encourage the recovery of the classical treasures of Pompeii and Naples.

The inventiveness of women remained an on-going gift to humanity. **Lady Mary Montagu** (1717) saw a new system for protecting against smallpox while visiting Turkey. She had suffered from the disease as a child. She brought the technique back to England, and smallpox inoculation was born – thus saving untold lives. It was the primary defense against the disease until the discovery of vaccination by Jenner 80 years later.[40]

Inventors

Catherine Littlefield Greene (1755 – 1814 CE) is a woman about whom there still is a bit of controversy. How much help did she give to Eli Whitney when the cotton gin was invented? We may never know. Whitney owns the patent, but apparently she suggested the idea and helped with the design. The patent shown (Figure 28) is from the Digital Library.[41]

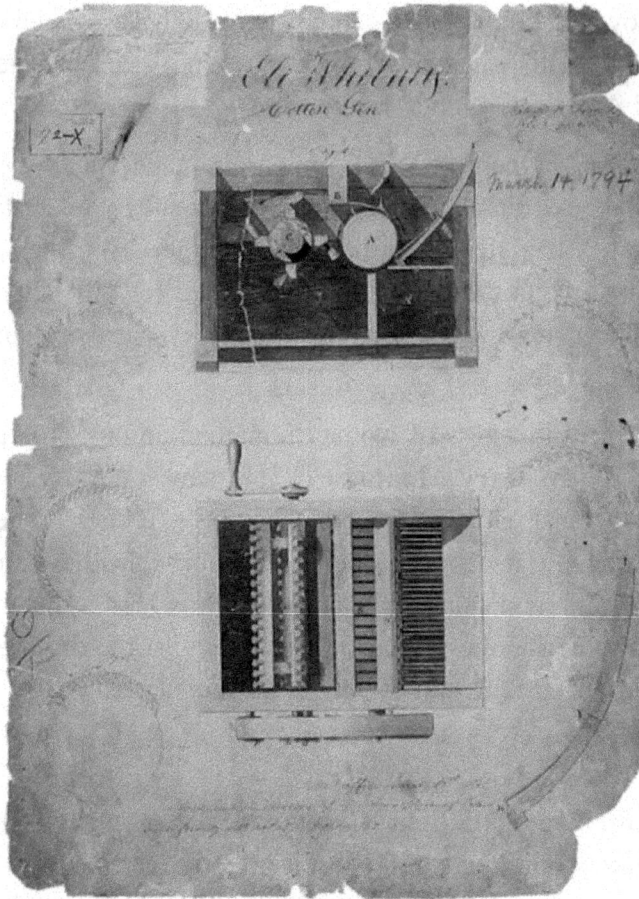

FIGURE 28 Patent for the cotton gin

The first woman to receive a patent *in her own name* in the United States was **Mary Dixon Kies** (1809) for an invention that allowed the easy weaving of straw with silk to make bonnets. In so doing she bolstered New England's hat economy, faltering because of an embargo on imported European goods. Straw bonnets manufactured in Massachusetts alone in 1810 had an estimated value of more than $500,000 or over $4.7 million in today's money. Dolly Madison honored her for this work. She was not the *very* first woman to receive a United States patent. That honor goes to **Mrs. Samuel Slater** in 1793 for cotton sewing thread.[42] But Mrs. Slater did not receive the patent in her own name; she received it in her husband's name. In 1815 **Mary Brush** received a patent for a corset. By 1851 fourteen patents had been awarded to women in the United States.[43] Among the articles invented were an ice-cream freezer (by **Nancy Johnson**), and a submarine telescope and lamp (**Sarah Mather** in 1845).

The Healing Arts

Finally the opportunity to obtain a medical degree was coming back to women. There is the interesting case of **Miranda Stuart**, alias Dr. James Barry (1795 – 1856). Her real surname was probably Bulkeley. She posed as a male to attend Edinburgh College, matriculating in 1812. Maintaining the charade, her career included an appointment as the Surgeon General of Canada. Her gender was not discovered until her death. Of course, in Italy women continued their long participation in medicine. **Maria Pettracini** (c. 1780 CE) and her daughter **Zaffira** were both teachers of anatomy at Ferrera. **Maria dalle Donne** (1778 – 1842 CE) obtained a doctorate from Bologna. Napoleon was so impressed by her that he instituted the chair of obstetrics in the university – a position she held until her death. Other countries now allowed women into medical schools. **Nadejda Suslowa**, a Russian, obtained an MD from the University of Zurich in 1867, the first woman to do so. In 1869 The Medico-Chirurgical Academy of St. Petersburg conferred an MD upon **Madame Kaschkarow**. Soon the doors opened again to women in medicine all over Europe and the United States.[44] There will be much more on this in the next Chapter. This is quite a change from the early 13th century when the university doors slammed shut in the face of women in France. The French even allowed female midwives to

127

practice medicine. **Marie Gillain Bolvin** (1773 – 1841 CE), a highly respected French midwife, received an honorary MD from the University of Marburg, Germany. Mozans[45] says that she would have been elected to the French Royal Academy had they accepted women. Her textbook on obstetrics was quite well received, and the King of Prussia awarded her the Order of Merit in 1814.

Mathematics and Astronomy

Mozans[46] mentions several learned women of France, although perhaps they are not as well known as their Italian sisters. His list includes Gilberte and Jacqueline Pascal (1625 – 1661 CE), Marie-Elionore de Rohan, Marie Cramoisy, Mlle. de Luynes, and Elisabeth de Rochechouart. Mme. Dacier (1624 – 1720 CE) was especially honored for her knowledge of Latin and Greek. Her version of the *Iliad* was translated into English.

Sophie Germain (1776 – 1831 CE), was a gifted mathematician who was called the Hypatia of the 19th century. She (Figure 29) taught herself basic mathematics, learning Latin and Greek, so she could read the important works by Sir Isaac Newton and the mathematician Leonhard Euler. Mozans tells us that her parents tried to prevent her studying at night; she would sneak down to the flickering fire and hide under a blanket so she could continue to read. Never formally trained, she was unable to attend school because of her gender.

She sent one of her papers to the mathematician Lagrange who was so overwhelmed by it that he offered himself as her mentor. She even impressed the rather gruff mathematician Karl Friedrich Gauss (one of the rare geniuses in mathematics) who said about her:

FIGURE 29 bust of Sophie Germain

"But, when a person of the sex, which according to our customs and prejudices, must encounter infinitely more difficulties than men to familiarize herself with these thorny researches, succeeds nevertheless in surmounting these obstacles and penetrating the most obscure parts of them, then without doubt she must have the noblest courage, quite extraordinary talents and a superior genius."

She proved that the first case of Fermat's last theorem is true for certain prime numbers – when both p and $2p + 1$ are prime, then p is called a Sophie Germain prime, and the theorem is true. The first few Sophie Germain primes are

$$2, 3, 5, 11, 23, 29, 41, 53, 83, 89, 113, 131 \ldots$$

She also made important contributions in the field of elasticity of metals. Why is this important? Because metals bend under stress and strain. Bridges, for example, are made of metal and can bend and break under the right kind of strong wind. The elasticity is a measure of how much the metal will bend under stress and strain. Her work fed directly into the work of Eiffel (who subsequently built the Eiffel Tower).

Of course we have wonderful astronomers of this time. The *real* Solar System was finally defined by the computation of precise orbits of the planets and comets (such as Halley's Comet). The Copernican Revolution had come into its own.

One famous practitioner was Sir William Herschel, a musician and astronomer. He discovered the planet Uranus,[47] made several star catalogs, and mapped out the Milky Way. To help with all this he had a faithful assistant, his sister, **Caroline Lucretia Herschel** (1750 – 1848 CE) whom he brought from their home in Hanover, Germany to work with him in England (see Figure 30). She was an accomplished astronomer in her own right, discovering eight, possibly nine, comets as well as several nebulae (galaxies) and clusters. She helped her brother in all his duties from telescope lens grinding to observing the stars at night. After the discovery of Uranus she received a government salary of £50 a year for her work making her, perhaps, the first woman to hold a government position in England. The two siblings made a remarkable team.

Together they transformed astronomy from the study of the clockwork universe of Newton to the exploration of a universe in which everything has a life story.

Their home in Bath, England is a now a museum. Although she was an indefatigable astronomer, she was not entirely pleased at the mess her brother's experiments made of their home. He built all the equipment they used. She wrote "*It was to my sorrow that I saw almost every room in the house turned into a workshop.*[48]"

FIGURE 30 Caroline Herschel

Caroline had a number of her works published including 'A catalogue of eight hundred and sixty stars observed by Flamsteed but not included in the British Catalogue' and 'A general index of reference to every observation of every star in the above-mentioned British Catalogue'. She received a gold medal in 1828 from the Royal Society for her work: The reduction and arrangement in the form of catalogue, in zones, of all the star-clusters and nebulae observed by Sir W. Herschel in his sweeps. These catalogs formed the basis for the modern catalogs used in astronomy today. She received two other gold medals – one from the King of Denmark and one from the King of Prussia. In 1835 she was elected, along with Mary Somerville, an honorary member of the Royal Astronomical Society. They were the first two women so honored. Three years later she was elected an honorary member of the Royal Irish Academy. Her epitaph, composed by herself, contains the following:

"The eyes of her who is glorified were here below turned to the starry heavens. Her own discoveries of comets and her participation in the immortal labours of her brother, William Herschel, bear witness of this to future ages. The Royal Academy of Dublin and the Royal Astronomical Society of London numbered her among their members. At the age of 97 years and 10 months she fell asleep in happy peace, and in full possession of her faculties; following to a better life her father, Isaac Herschel, who lived to the age of 60 years 7 months and lies buried near this spot since the 25th March, 1767. "[49]

Her fellow astronomer-mathematician, **Mary Somerville** (1780 – 1872 CE), was likewise honored in her own time. Mary translated with commentary the difficult work by mathematician/astronomer Laplace – *Mécanique Céleste* – into English, an effort for which her English contemporaries owed her much. It was used as a college textbook for nearly a century. Laplace himself said that she was the only woman to understand his work. She was mainly self-taught. She produced an amazing series of papers, presented on her behalf to the Royal Society by her husband. After her death Oxford University named one of its colleges after her. She had the incredible gift of synthesizing the works of others into comprehensible language. She wrote a book called *The Connection of the Physical Sciences* in which she discusses the possibility of an undiscovered planet being responsible for the irregularities in Uranus's orbit. This led directly into John Couch Adams' research at Cambridge on the possibility of another planet, and subsequent discovery of the planet Neptune. Ogilvie's book *Women in Science*[50] has a very nice article about her as does the web site at Agnes Scott College.

Another astronomer of the time was **Marie Jeanne de Lalande** (c. 1790 CE). She assisted her husband and cousin in their astronomical work. She certainly was good at detailed and exacting computations because her work appeared in the *Connaissance des Temps* – an astronomical almanac still produced by the French. These almanacs are detailed listings of future positions of the Sun, Moon, and planets versus time. It takes incredibly laborious computations and complicated celestial mechanics to produce these almanacs. Only a few nations have astronomers gifted enough to produce them. Paris Observatory often hired women as "computers" – people to do the detailed computations needed to produce the almanacs.

We are privileged to have an astronomer from China, **Wang Zhenyi** (1768 – 1797 CE). She studied lunar eclipses using models she constructed herself (Figure 31). She wrote twelve books on astronomy and mathematics. She derived the precession of the equinoxes. She wrote that men and women "are all people, who have the same reason for studying". Her poetry depicted the hard life of the poor. She spent years gathering data on the heavenly bodies and the clouds. The data gathered from her attempts to measure atmospheric humidity were used for weather forecasting and, purportedly, could predict floods and droughts.

The reputation of the United States in science, once thought to be the result of small efforts by a bunch of rag-tag amateurs, grew to world class status during the 19th and early 20th century. So as we move forward in time we begin to find many superb women of science in the United States. No longer did American students of science have to travel to Europe to receive training.

They could turn their eyes "to the starry heavens" with their European sisters.

FIGURE 31 Wang Zhenyi

17th century	Marie Guyart, Marie de l'Incarnation
18th century	Jane Haldimand Marcet
	Claudine Pouliet Picardet
	Jane Colden
	Martha Daniell Logan
	Agnes Ibbetson
	Martha Laurens Ramsey
	Maria Angela Ardinghelli
	Jane and Mary Parminster
	Margaret Bryan
	Lucinda Foote
	Mary Lyon
	Marie Anne Pierrette Lavoisier
	Elizabeth Fulhame
	Caroline
	Lady Mary Montagu
	Catherine Littlefield Greene
	Mrs. Samuel Slater
	Maria Pettracini
	Zaffira
	Caroline Lucretia Herschel
	Marie Jeanne de Lalande
19th century	Margaretta Hare Morris
	James Barry
	Josephine Kablick
	Jeanne Villepreux-Power
	Mary Anning
	Emma Willard
	Mrs. Mary Norris
	Almira Hart Lincoln Phelps

Jane Andrews
Mary Dixon Kies
Mary Brush
Nancy Johnson
Sarah Mather
Maria dalle Donne
Miranda Stuart
Nadejda Suslowa
Madame Kaschkarow
Marie Gillain Bolvin
Sophie Germain
Mary Somerville
Harriet Morrison Irwin
Elizabeth Isis Pogson Kent
Mary Orr Evershed
Wang Zhenyi

[1] Caroline Herschel

[2] *The Alphabet Versus the Goddess*, Leonard Shlain, Viking, 1998

[3] Charles F. Horne, *The Sacred Books and Early Literature of the East* (New York: Parke, Austin, & Lipscomb, 1917), Vol. II: Egypt, pp. 62-78.

[4] Old Assyrian, 19th century BCE Text: B. Hrozný, *Inscriptions Cunéiformes du Kultépé* (Praha, 1952). Transliteration and translation, Hrozný, in *Symbolae Koschaker* (*Studia et Documenta* II, 1939), 108*ff.*

[5] A woman sacred to the temple

[6] Infertility was seen as a woman's issue (fault) throughout the ancient world.

[7] The Vedas are India's ancient sacred texts.

[8] Manu Smriti III.55-59 – from the Vedas

[9] Athravaveda Samhita, Part 2, Kanda 27, sukta 107, sloka 5705

[10] Nadītu or Naditu is the designation of a legal position for women in Babylonian society. Nadītu lived in monastic buildings, but in general did own their home, and were independent. They could engage in contracts, borrow money, and perform other business transactions normally denied to women. Usually these women were part of the elite, often from royal families.

[11] For example, an adulterer in Mosaic Law was killed (both the man and the woman). In Babylonian Law the woman could be forgiven by her husband if he wished.

[12] *The Hittites*, J. Lehman, Viking Press, 1977

[13] Introduction to the Cataloque of the "Woman in Anatolia, 9000 years of the Anatolian Women Exhibition" by Prof. Günsel Renda (Ministry of Culture Publication, İstanbul, 1993)

[14] The Amazons were said to be from the Anatolian plain.

[15] Pomeroy, Sarah B. *Goddesses, Whores, Wives, and Slaves*, Schocken Books, 1975

[16] The term refers to the time when an unusually large number of great philosophers lived: *e.g.*, Plato, Socrates, the Buddha, Lao Tsu, Confucius, Mahavira, Jeremiah.

[17] Pomeroy, Sarah B. *Goddesses, Whores, Wives, and Slaves*, Schocken Books, 1975, 3

[18] Hellenistic Greece begins with the conquests of Alexander which spread much of the Grecian philosophy throughout a wide region.

[19] It was at the time that the city of Alexandria was constructed and the Great Library begun.

[20] Loeb Classical Library, *Select Papyri I*, Harvard University Press, 1932

[21] Pomeroy, Sarah B. *Goddesses, Whores, Wives, and Slaves*, Schocken Books, 1975

[22] *The Alphabet Versus the Goddess*, Leonard Shlain, Viking, 1998

[23] The biological study of the functions of living organisms and their parts

[24] The microscope was invented in 1683.

[25] *The Life and Times of Martha Laurens Ramsay*, 1759 –1811[25] is a book by Joanna Bowen Gillespie.

[26] The first important work in vegetable physiology

[27] Observations and physical experiences of several marine and terrestrial animals

[28] http://www.hevac-heritage.org/items_of_interest/heating/national_trust_properties/a_la_ronde/a_la_ronde.htm

[29] Darwin's *On the Origin of Species* was published in 1859.

[30] There is an excellent web site about her:

www.discoveringfossils.co.uk/Mary_Anning.htm

[31] the widow of the former Recorder of the City of London

[32] Annotation on an undated letter from Mary Anning to one of the Misses Philpot of Lyme, in the collection of the American Philosophical Society, Philadelphia, cited in Torrens, Hugh: Mary Anning (1799-1847] of Lyme: 'the greatest fossilist the world ever knew,' *British Journal for the History of Science*, **25**: 257-84, 1995.

[33] http://www.scienceandsociety.co.uk/ is a link to a portrait of her with her children

[34] *Women in Science*, Ogilvie, MIT Press, 1986

[35] *Women on the Margins*, Davis, Harvard College, 1995

[36] *The book of Women*, Lynne Griffen and Kelly McCann, 1992, Bob Adams, Inc.

[37] http://web.ukonline.co.uk/m.gratton/index.html is a wonderful web site listing women's firsts.

[38] a title he shares with Robert Boyle

[39] *Women in Chemistry: Their Changing Roles from Alchemical Times to the Mid-Twentieth Century.* Marelene and Geoffrey Rayner-Canham, American Chemical Society and the Chemical Heritage Foundation, Washington, DC, 1998

[40] http://www.foundersofscience.net/lady_mary_montagu.htm contains a letter from her describing the technique.

[41] http://www.digitalhistory.uh.edu/daybyday/daybyday.cfm?db=abolition

[42] *Mothers of Invention*, Vare and Ptacek, William Morrow & Company, 1988

[43] By 1900 over 3500 patents had been awarded to women in the United States.

[44] Medical and dental schools in the United States are now routinely 50% female.

[45] *Woman in Science*, Mozans, 1913, D. Appleton & Company

[46] ibid

[47] The first planet discovered since ancient times.

[48] booklet from the William Herschel Museum

[49] *Woman in Science*, Mozans, 1913, D. Appleton & Company

[50] MIT Press, 1986

CHAPTER 5

THE 19th AND EARLY 20th CENTURIES

"we especially need imagination in science"[1]

𝕿he 19th century became a century of firsts for women. Many women were able to obtain the advanced degree of doctor of philosophy, in some cases the first at their schools to do so. Whereas in earlier centuries we looked for scholars, philosophers, inventors, poets, and others without academic degrees, now we can also look for scholars with degrees.

A PhD is a <u>D</u>octor of <u>PH</u>ilosophy degree, the highest academic honor a university can bestow. The 'doctoral degree' originated in the ninth century schools of the Muslim world before spreading to European universities. Originally awarded in the professions of law, medicine, and theology, the PhD eventually became the designation for doctoral degrees in disciplines outside of these fields. The first PhD was awarded in Paris in 1150, but it was the early 1800s before the degree gained its contemporary status.

One of the qualifications for the degree is that the recipient must have accomplished some work that is creative and new – never done before. Typically the PhD candidate goes before a board of examiners, who question and challenge the candidate on her work. If the candidate passes this final test, then the degree is awarded. It is now the "scientist", who was the "natural philosopher".

In the United States the PhD indicates the level of training the scientist has. There are other levels. A bachelor's degree comes typically after four years at a college or university. A master's degree is advanced training beyond the bachelor's degree, sometimes as little as one year more, sometimes more than that. A medical degree, MD, also takes advanced training beyond the bachelor's degree as does a law degree. And then the PhD is the most advanced academic degree.

As usual, Italy continued to lead the way. The Italians kept pouring out learned women. Doctorates of civil law from the University of Bologna went to **Maddalena Canedi-Noe** (1870) and **Maria Vittoria Dosi**. And now I shall leave the Italians, knowing that they continue to produce scholarly women, and move to other countries. In the United States women began receiving bachelor's degrees as early as 1840 and advanced degrees as early as 1853 (see Maria Mitchell later in this chapter). The PhD's came a bit later[2]. By 1891 more than eighty-two percent of all the

142

American public school teachers were women; over two hundred colleges had over four thousand women students; industrial schools for girls had been established in almost every state.[3]

This chapter will concentrate on the United States and Europe. In fact the names are so numerous that I shall list them in almost a dictionary format and leave it to the reader to find out more about them. This is a straightforward task with the Internet. It is wonderful that the names are numerous, but I am sure that many more women are there to be found. Limiting the selection to women mostly from the US and Europe is a choice mandated by the sheer numbers of technical women found there and by the lack of easily accessible information on women elsewhere. Even so, this is the longest chapter in the book.

Some women, when denied by schools in the United States, went to Europe. An example was **Martha Carey Thomas** (1857 – 1935 CE) who was denied a PhD by universities in the United States, (she graduated from Cornell in 1877) so she went to Europe, and in 1882 received the first PhD in linguistics given to a woman by the University of Zurich, Switzerland. She returned to the United States to assume the position of Dean at Bryn Mawr, the first women in the United States to hold the title of Dean. Bryn Mawr College opened its doors in 1885 and was the first to offer graduate education through the PhD — a signal that its founders refused to accept the limitations imposed on women's intellectual achievements by other institutions. Dr. Thomas said

"A woman can be a woman and a true one without having all her time engrossed by dress and society."

Ah-ha! One can think of science, technology, and scholarship as well as dress and society and still be a proper woman. It is a freeing thought.

The Healing Arts

This century led to a number of firsts for women in science but it was also the century of the American Civil War that decimated the southern states. The push for needed medical care for the survivors emboldened women to study medicine.

Starting off, therefore, with medicine we find that women continued to flourish as they forced open the doors to medical schools. It was especially important for women to enter the field because women still suffered from physicians' mistaken belief that all their medical woes stemmed from the womb. Many physicians mistreated their patients by ignoring the obvious and treating women as nothing but a reproductive unit. It took a strong woman to overcome the general prejudice and gain the treatment she needed.

The 19th century saw the growth of medicine into almost a true science. In the last decades of the 19th century European medicine became far more powerful by allying itself with emerging scientific techniques and technologies. Germ theory, bacteriology, and chemical analysis joined new diagnostic instruments and drug therapies in changing clinical practice. At the same time, doctors and surgeons fought to establish themselves as professional groups. Medical schools grew in importance. In 1858 Britain passed the Medical Registration Act which required that physicians be educated and take exams. Prior to this time over a third of physicians in Britain had little to no official training, even though medical schools did exist. Today, of course, a physician is licensed by the state to practice medicine only after extensive training.

Women quickly made their mark in professional medicine. The mid-nineteenth century saw several firsts. Seven years to the day after Maria dalle Donne's (Chapter 4) death **Elizabeth Blackwell** (1821 – 1910 CE) decided to enter college to study medicine and surgery (Figure 32).

Although born in England, she moved to the United States as a child and finally succeeded in her medical goal at Geneva College in Geneva, New York where she received the first MD given to a woman in the United States (January 23, 1849 CE). The web site[4] of the US National Library of Medicine maintained by the National Institutes of Health (NIH) has a nice article about her life.

FIGURE 32 US stamp of Dr. Blackwell

The number of women in medicine in the United States increased dramatically with the opening of the New England Female Medical College in Boston on November 1, 1848 to train midwives. By 1850 it began training women to receive the MD.

Dr. Samuel Gregory and Dr. Israel Talbot headed the College. When condemned by the male community for training "weak" women Dr. Gregory responded

"Suppose physicians were as ignorant upon this subject as females now are; they would then be easily alarmed and incapable of rendering efficient and in case of emergency ... the fact of being one of the stronger sex does not render one competent."

One of the early teachers there was **Marie Zakrzewska**, (1829 – 1902 CE) a German-born pioneer of women in medicine who received her medical degree from Cleveland Western Reserve College in 1856. A year later in 1857 **Esther Hawks** (1833 – 1906 CE) graduated from the New England Female Medical College and shortly afterward became a physician. She had attended public schools in New Hampshire, becoming a teacher. After marriage she taught a school for black children south of Tampa, Florida. This was illegal at the time but no complaint was lodged. Returning to the north she received her medical degree while actively supporting abolition. Returning to the south during the Civil War years she practiced medicine. She also ran the first racially integrated school in the south. You can read the story of her life in her diary.[5]

The sole American female MD in 1849 led to about 200 MDs by 1860.[6] A mere twenty years after that, the census of 1880 showed that there were 2,400 women of medicine. By the end of the 19th century in 1900 the number of women with MDs had increased to over 7000, and women were well on their way. Rather quickly they became leaders in the field.

Dr. Blackwell was not the first woman in the United States to practice as a doctor though. That honor goes to **Harriet Hunt** (1805 – 1875 CE) who set up shop in Boston in 1835. She had received her medical training, as many did in those days, through apprenticeship. In 1847 Harriet became the first woman to apply to Harvard Medical School[7]. She was rejected. In 1853 Harriet was finally awarded an honorary degree from the New England Female Medical College. She noted:

"The prevailing idea ... is, that the doctor is to cure the disease. It is not so. The doctor and the patient together, are to cure or mitigate the disease. They must be coworkers....."[8]

which is a sentiment still worthy today.

The second woman to receive an MD in the United States was **Lydia Folger Fowler** (1822 – 1879 CE) who received the degree in 1850 from Central Medical College, Syracuse, New York, which was the first medical institution to admit women on a regular basis. She became the first female professor in an American medical college. On March 11, 1850, the Pennsylvania legislature passed an act to incorporate the Female Medical College of Pennsylvania – one of the first regular medical schools for women in America. Another first was achieved by **Sarah Read Adamson Dolley** (1829 – 1909 CE) who became the first women to intern in a hospital (1851). She graduated from Central Medical College. The web site at NIH has a nice biography of her.[9]

The Claypole twins were remarkable achievers. **Edith Jane Claypole** (1870 – 1915 CE) was a physiologist and pathologist. Her identical twin sister **Agnes Mary Claypole** (1870 – 1954 CE) was a zoologist. Both attended Buchtel College (now the University of Akron) in Akron, Ohio, where their father was teaching at the time, and graduated in 1892. Both entered graduate school at Cornell. Edith earned a master's degree in 1893 with a thesis on the blood cells of amphibians; Agnes received a master's degree in 1894 with a thesis on the digestive tract of eels. Edith then went to teach at Wellesley College (which opened in 1875), while Agnes went to the University of Chicago, where she earned her PhD in 1896. After two years at Wellesley, Edith entered the Cornell Medical School and then completed her medical degree at the University of California, Los Angeles. Agnes also moved to California where she became the first female professor at the Throop Institute – now known as the California Institute of Technology. The 1902-1903 catalog lists Agnes Mary Claypole as Instructor in Zoology, and Edith Jane Claypole as Instructor in Biology and Bacteriology. Edith left the faculty to complete her medical education. In 1903-4 Agnes became Professor of Natural Science and Curator. She left after holding the position for only one year, however – she married Robert O. Moody, a professor of anatomy at U.C., Berkeley in 1903, and afterward moved to northern California (eventually joining the faculty at Mills College).

Elizabeth Garrett Anderson (1836 – 1917 CE) was a British physician who took an especially clever route to her degree. She discovered that the Society of Apothecaries did not specify that females were banned from taking their examinations. In 1865 Garrett sat for and passed the Apothecaries examination. As soon as she was granted the certificate that enabled her to become a doctor, the Society of Apothecaries changed their regulations to stop other women from entering the profession in this way. In 1870 she sat for her medical examinations at the University of Paris in France, passing them with ease. The British medical registry refused to recognize her degree. Nevertheless in 1872 she opened the New Hospital for Women in London.

In 1864 **Dr. Rebecca Lee Crumpler** (1831 – 1895 CE) became the first African-American woman to graduate from medical school (New England Female Medical College). She worked in the post-Civil War South and then in Boston and secured her well-earned place in the historical record with her book of medical advice, *Book of Medical Discourses*, for women and children, published in 1883.[10]

Mary E. Britton (1855 – 1925 CE) was the first African-American physician to graduate from Berea College in Kentucky and the first African-American woman (see Figure 33) to practice medicine in Lexington, Kentucky.[11] She specialized in hydrotherapy and electrotherapy. Both she and Dr. Crumpler must have been incredibly brave, considering the times.

(Courtesy of the Berea Special Collections and Archives)

FIGURE 33 Dr. Mary E. Britton
Courtesy Berea archives

Mary Walker (1832 – 1919 CE) was a physician (graduated in 1855 from Central Medical College in Syracuse) who dressed as a man during the American Civil War to perform the duties of a physician and surgeon. She worked first as a nurse in the improvised hospital in the U.S. Patent Office and then as a physician and surgeon. For her service she was awarded the Congressional Medal of Honor only to have it later revoked in

1917 (she was ridiculed for many of her ideas and assertive manner). She refused to return the Medal. The honor was restored to her in 1977 by President Jimmy Carter. A 20¢ stamp honoring Dr. Mary Walker was issued in Oswego, NY on June 10, 1982 (Figure 34). She is the only woman to receive the Congressional Medal of Honor.[12]

Dentistry was not neglected by women. **Lucy Hobbs Taylor** (1833 – 1910 CE) may have been the first women in the world to receive the degree of Doctor of Dental Surgery. She graduated from Ohio Dental College in 1866. A year later she and her husband, also a dentist, opened a joint office and built up the largest dental practice in the city. Her home is still there in downtown Lawrence, Kansas. In 1890 **Dr. Ida Gray** (1867 – 1953 CE) became the first African-American woman to become a dentist in the United States. She graduated from the University of Michigan.

FIGURE 34 US stamp featuring Dr. Walker

Her commendation for the medal reads

Whereas it appears from official reports that Dr. Mary E. Walker, a graduate of medicine, "has rendered valuable service to the government, and her efforts have been earnest and untiring in a variety of ways," and that she was assigned to duty and served as an assistant surgeon in charge of female prisoners at Louisville, KY., under the recommendation of Major-Generals Sherman and Thomas, and faithfully served as contract surgeon in the service of the United states, and has devoted herself with much patriotic zeal to the sick and wounded soldiers, both in the field and hospitals, to the detriment of her own health, and has endured hardships as a prisoner of war four months in a southern prison while acting as contract surgeon; and

Whereas by reason of her not being a commissioned officer in the military service a brevet or honorary rank can not, under existing laws, be conferred upon her; and

Whereas in the opinion of the President an honorable recognition of her services and suffers should be made;

It is Ordered. That a testimonial thereof shall be hereby made and given to the said Dr. Mary E. Walker, and that the usual medal of honor for meritorious services be given her.

Given under my hand in the city of Washington, D. C. this 11th day of November, A.D. 1865.

Andrew Johnson, President

By the President: Edwin M. Stanton, Secretary of War

Josephine S. Baker (1873 – 1945 CE) was a physician and health researcher. She went to the Women's Medical College in New York City to earn her MD. She said:

"It is true that the laboratory and the X-ray have added much that is valuable to our knowledge of diagnosis, but in this change of tactics the average doctor has lost much of his basic skill. Thirty years ago, we had to depend upon our sense of touch, sight, and hearing to make a diagnosis, and experience developed alertness that is not completely replaced by routine laboratory reports."

She took a part-time job with the Department of Public Health as a public school health inspector where she rose in the ranks to create and run the Bureau of Child Hygiene. She helped track down "typhoid Mary" (Mary Mallon[13]). She was one of the first doctors to recognize how important being held is to a child's health. She perfected the application of sliver nitrate eye drops to infants, now a standard procedure to prevent eye infections in newborns. She was the first woman to be assistant surgeon general in the United States. She was also the first woman representative to the League of Nations – as Health Committee representative for the United States.

Florence Rena Sabin (1871 – 1953 CE) was an anatomist and histologist (study of tissue sectioned as a thin slice) who trained at Johns Hopkins Medical College where she received an MD in 1900. She was the first woman appointed to the faculty at Johns Hopkins University. In 1925 she became the first woman elected to the National Academy of Sciences.[14] She was one of the first women to establish a medical research career and was proclaimed the greatest scientist of our time by the Rockefeller Institute. There is a nice article on her in the online *Encyclopedia Britannica.* She was co-identifier of the cell that typifies Hodgkin's disease (a cancer that starts in lymphatic tissue).

Mary Putnam Jacobi (1842 – 1906 CE) was a physician who, in 1876, won the Harvard Bolyston Prize for her essay on "the effects of rest on menstruation". In 1864 she studied at the Medical College of Pennsylvania. She then went to France where she studied at the Ecole de Médecine. She received her degree from there in 1871 (the second woman to do so). She was elected to the National Women's Hall of Fame.

152

Nettie Maria Stevens (1861 – 1912 CE) was a cytogeneticist (the study of chromosomes from samples of bodily fluids). She received her PhD from Bryn Mawr in 1903 and then studied in Europe. She is the discoverer of the chromosomal determination of sex (those X and Y chromosomes that determine whether the baby is a boy or girl – see Figure 35) and published about thirty-eight professional papers.

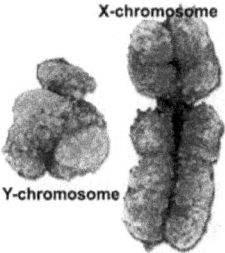

FIGURE 35

Frances Elisabeth Crowell (c 1910) was a nurse who specialized in tuberculosis. She was appointed executive secretary of the Board of Tuberculosis Clinics in 1910 from which she helped launch the international public health movement for which she received the Legion of Honor from France.

Alice Hamilton (c. 1900) received an MD from the University of Michigan. Her work was a model for the study of occupational disease. She became one of the leading toxicologists in the US.

And more firsts came their way. **Alice Bennett** (1851 – 1925 CE) was probably the first woman to be a superintendent at a state hospital for the mentally ill (Norristown, Pennsylvania) where she introduced the concept of occupational therapy for the patients. After getting the MD she earned a PhD in anatomy from the University of Pennsylvania in 1880, the first woman to get a PhD from this university. There are a few people with both an MD and a PhD but not many, even today. **Anita Newcomb McGee** (1864 – 1940 CE) was the founder of the Army Nurse Corps and the first women appointed assistant surgeon general in the US Army (1898). She received her MD from George Washington University.

The firsts kept coming. In 1876 **Sarah Stevenson** (1841 – 1909 CE) was the first woman to become a member of the American Medical Association (AMA). She studied medicine at the Woman's Hospital Medical College in Chicago. After a year, she went to London for a year where she studied with Thomas Huxley at the South Kensington Science School. Returning to Chicago and the Woman's Hospital Medical College, she graduated with an MD in 1874. She began a private practice in Chicago and published *Boys and Girls in Biology* for high school students, based in part on Huxley's lectures that she had attended in London. In 1876 she attended the AMA convention as a delegate of the Illinois State Medical Society. Her presence was accepted without significant challenge, and she became the AMA's first female member.

In 1896 **Shih Mai-Yu** became the first Chinese woman to obtain an MD from an American university (the University of Michigan). A crater on the planet Venus is named for her.[15] While a graduate student at Cornell University **Martha Tracy** (1876 – 1942 CE) assisted William Coley to develop Coley's fluid, used in the treatment of sarcoma (malignant tumors of the connective tissues). **Mary Harris Thompson** (1829 – 1895 CE) was a surgeon who helped to establish the Chicago Hospital for Women and Children. She was the first, and for many years, the only woman to perform major surgery in Chicago. Another influential early doctor was **Martha G. Ripley** (1843 – 1912 CE),

154

who attended the Boston University Medical School in 1879. She began her practice in Minneapolis in 1883. Maternity Hospital was developed from a clinic opened in 1886 by Dr. Ripley in a small Minneapolis house.

Lydia Adams Dewitt (1859 – 1928 CE) became an associate professor of pathology at the University of Chicago. She received her MD in 1898. A pathologist and research scientist known for her contributions to the anatomy of the pancreas and heart and for pioneering work in the chemotherapy of tuberculosis, she founded the Woman's Research Club at University of Michigan. Lydia Adams DeWitt Research Awards are available at the University of Michigan to help meet career-relevant needs of individual faculty on the research track that will increase the participation and advancement of women faculty. DeWitt awards support a range of activities necessary for scholarly work in science and engineering fields. Competition for funds ($40,000 annually) takes place once each year. Applications may be for up to $20,000.

FIGURE 36 Order of Merit

One cannot ignore **Florence Nightingale** (1820 – 1910 CE) and the field of nursing. Not only was she a nurse of international reputation she was also a statistician. In 1860 for her contribution to Army statistics and comparative hospital statistics Florence Nightingale became the first woman elected a fellow of the Statistical Society. Her writings on hospital planning and organization had a profound effect in England and across the world. She was the first person in the Western world to introduce statistics into public health. She also introduced the concept of the pie chart so useful in making presentations! In 1907 she became the first woman to receive the British Order of Merit (Figure 36). She lectured on many topics including the sad state of rural hygiene in England at the turn of the century.

Sister Kenny (c. 1886 – 1952 CE) was a pioneer in the treatment of polio before the discovery of the Salk vaccine. Her autobiography (written in 1943) entitled *And They Shall Walk* was made into a movie. She trained as a nurse in the Australian medical corps and served during World War I. After the War she learned more about polio and developed her own ideas for treatment. She invented a special stretcher to transport patients in shock. Royalties from the patent gave her the money to start her own clinic for the treatment of polio. She advocated applying heat and physical therapy to polio victims in opposition to the medical establishment that advocated immobilization. She came to the United States to the University of Minnesota medical center. Although her methods were never formally endorsed by the medical profession she got good results. In the late 1950's the Kenny Institute and the World Health Organization were the major supporters of continued polio research. The Kenny Institute survives as part of the Abbott Northwestern Hospital in Minneapolis.

Linda Richards (1841 – 1930 CE) became the first registered nurse in the United States after enrolling for training at the New England Hospital for Women and Children. She received her diploma in 1873. She also worked for five years in Japan where she started a training school for nurses. She became the first president of the American Society of Superintendents of Training Schools. She was inducted into the National Women's Hall of Fame (http://www.greatwomen.org/). There are only 207 entries in the Hall.

Pioneer researcher in the scientific aspects of nutrition, **Mary Davies Swartz Rose** (1874 – 1941 CE) in 1912 published a widely used book: *A Laboratory Handbook for Dietitians* and in 1927 published *The Foundations of Nutrition* still found in libraries.

The Right to Vote

Moving on from medicine, a major world-wide achievement for women occurred when they got the vote, giving them more of a chance to affect their own lives. Typically women obtained local voting rights before they received national rights. For example, in 1869 Britain granted unmarried women who were householders the right to vote in local elections. In 1893 New Zealand granted women the right to vote. Women got full rights to vote in 1918 in England, Germany, Latvia, Poland, Estonia, and Russia, followed two years later in 1920 by the United States. It was not until 1968 that Switzerland granted women the right to vote in Federal elections. Kuwait granted women the right to vote in 2005. They are not allowed to vote in Saudi Arabia. Most countries, however, now allow women this basic right of citizenship.

Women's rights were addressed head on in the 19th century. There were several conferences on women's rights. Sojourner Truth (1797 – 1883 CE) gave her famous "Ain't I a Woman" speech in 1851 at the Women's Convention in Akron, Ohio. John Stuart Mill (1806 – 1873 CE) used a lot more words to say essentially the same thing in his *Subjection of Women* essay. Others were Mary Wollstonecraft (*Vindication of the Rights of Women*) and Susan B. Anthony (1820 – 1906 CE) (*On Women's Right to Vote*). Although women got the right to vote in the United States in 1920, it was not until 1938 that the Fair Labor Standards Act set limits to child labor in the United States. The United States has yet to ratify the United Nations Convention on the Rights of the Child. Although condemned by the United Nations, according to UNICEF over 246 million children world-wide are forced to work, often in horrific conditions. Freedom to engage in scholarly endeavors is not yet part of the world-wide human condition.

Biology and Natural History

Marilyn Baily Ogilvie has written an excellent book titled *Women in Science*.[16] In it she lists a number of technical women in the 19th and early 20th century. Many of the women she lists are botanists, natural historians, or in related fields. Rather than duplicate all her information I shall simply briefly mention some of these women and refer the reader to her book.

Women had stayed the course in astronomy and mathematics, and they were now appearing in the other sciences as well. The study of plants continued to capture the imagination of many women in the 19th century. And of course many of the women went far beyond the recording of the natural world, they also experimented, and they were intrepid travelers. Many also achieved scholarly degrees in the various fields of biology. It was this century that biology saw its ground-breaking work in the 1859 publication of Darwin's *On the Origin of Species.*

Gulielma Lister (1860 – 1949 CE) was the niece of the famous surgeon Lord Lister (the mouth cleanser Listerine is named for him). She and her father were amateur naturalists as well as accomplished artists. Their work on fungi in particular was important in the classification of these plants. She was also an expert on the slime mold; unrelated to fungi, slime molds are enormous single cells that make them extraordinarily useful for studying cellular interactions. Trained at home by her father, Gulielma became a founding member of the British Mycological Society (president in 1912 and 1932). She was a fellow, a council member, and a vice-president of the Linnaean Society.

Mary Lua Adelia Davis Treat (1830 – 1923 CE) was an economic entomologist, botanist, and supporter of the early theory of evolution. She corresponded with Charles Darwin and supported herself by writing popular level science and collecting plants and insects for other researchers. She published her first article at age thirty-nine. In the next twenty-eight years she wrote seventy-six scientific and popular articles and five books. Darwin wrote that her experiments on controlling the sexes of butterflies were by far the best which have ever been made and asked her to repeat experiments on the relation of sexes of butterflies to the nutrition of the larvae. Her descriptions of insect behavior helped taxonomists to classify new species. She herself discovered a new species of orange aphid, an *Ichneumonid* fly (there are over 3,300 species of this type in the United States), two spiders, an amaryllis, and an ant (the ant, *Aphanogaster treatiae,* and the amaryllis, *Zephyranthus treatiae,* were named in her honor).

Cornelia Clapp (1849 – 1934 CE) studied at Mt. Holyoke majoring in zoology. She, after receiving a PhD from the University of Chicago in 1896, became a professor of zoology at Mt. Holyoke College and was active in the new (1888) Marine Biology Laboratory at Woods Hole, Massachusetts. This Laboratory still exists as one of the premier marine biology laboratories in the world. Another American zoologist was **Jennie Arms Sheldon** (1852 – 1938 CE) who trained in Boston. Her work was mostly at the popular level. She was listed in *Who's Who in Women*[17] of 1914 as was **Emily Ray Gregory** (born 1863 CE). Emily Ray was an American zoologist with a PhD from the University of Chicago. Her book *Observations on the development of excretory system in turtles* is still available. And the similarly named **Emily Gregory** (1841 – 1897 CE) was a botanist with a PhD from the University of Zurich (1886), the first American to earn one in botany. Her doctoral thesis was titled "Comparative Anatomy of the Fitz-like Hair Covering of Leaf Organs". She accepted the first post-doctoral fellowship ever awarded to a woman at the University of Pennsylvania. Afterward she was unable to obtain a paid position at Bernard College so she taught there unpaid. When she was finally awarded a paid professorship there, she taught for only two years before her death.

Unfortunately many women had to accept unpaid positions in order to engage in their science. Even today a married woman may have to accept a position secondary to that of her husband. We call this the two-body problem in science. It remains an active issue. But in recognition of the importance Professor Gregory attached to cultivating young minds, Barnard students annually bestow the Emily Gregory Award for Excellence in Teaching to a member of the faculty.

Mary Jane Rathbun (1860 – 1943 CE) was a marine zoologist who received a PhD from George Washington University in 1917 and served as a staff member and then assistant curator at the US National Museum (now the Smithsonian) in Washington DC. She was known for establishing the basic taxonomic information on *Crustacea*. She built the reputation of the Museum's Division of Marine Invertebrates to its present high standard of excellence.

Margaret Clay Ferguson (1863 – 1951 CE) was a botanist who, in 1901, received a PhD from Cornell University. She studied chemistry and botany at Wellesley College and then returned to head the botany department there. She was the first woman to be president of the Botanical Society of America.

Graceanna Lewis (1821 – 1912 CE) was an ornithologist and teacher of astronomy and botany in Philadelphia and Clifton Springs, New York. She wrote and illustrated *A Natural History of Birds* in 1844. In 1862 she met John Cassin, the foremost ornithologist in America. Cassin soon became her friend and mentor, and she studied with him until his death in 1869. By the time her apprenticeship with Cassin was complete, she was the most educated naturalist in America. She also was an active abolitionist, and hundreds of escaped slaves made their way through the Underground Railroad using her farm as one of the stations along the way. **Ethel Sargent** (1863 – 1918 CE) was an English botanist who trained at Cambridge University. Her papers are archived at Girton College, Cambridge. **Elizabeth Gifford Peckham** (1854 – 1940 CE) was an American arachnologist (studies spiders) and entomologist (studies bugs) who received a PhD from Cornell University in 1916. She was listed, as many of these women were, in the *Who's Who of Women* of 1914.

FIGURE 37 Hydra

Florence Peebles (1874 – 1956 CE) was a biologist who received a PhD from Bryn Mawr College in 1900. She also received degrees from Woman's College of Baltimore (1895), the University of Halle, the University of Wurzburg, and the University of Freiburg. She began her academic career at Bryn Mawr College as an instructor of biology. She also worked at Tulane University, Lawrence and Clark, and California Christian College during her career. She also spent time at the Marine Biological Laboratory at Woods Hole, Massachusetts. She is known for her work on how external influences affect the development of living tissue; she studied regeneration in lower organisms, in particular the hydra (Figure 37), a multi-cellular organism.

Susanna Phelps (1857 – 1915 CE) was an embryologist and comparative anatomist with a PhD (1880) from Cornell University. Sigma Xi[18] elected women to full membership beginning in 1888. Five women were inducted into the Cornell chapter at this time, including Susanna Phelps. Anna Botsford Comstock, Harriet Marx, Julia Warner Snow, and Mary Wardwell were also inducted.

Edith Patch (1876 – 1954 CE) was an entomologist, an aphid specialist (Figure 38), who received her PhD from Cornell University in 1911. In 1930 she became the first woman to be president of the Entomological Association of American and also president of the American Nature Study Society. One new genus and several species were named after her. She worked at the University of Maine which has named a building after her.

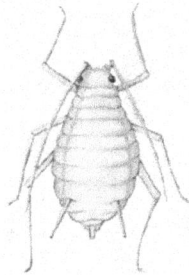

FIGURE 38 Green peach aphid

Mary Murtfeldt (1848 – 1913 CE) was an entomologist trained at Rockford College, Rockford, Illinois. She made significant contributions to the understanding of how yucca plants are pollinated and was listed in *Who's Who of Women* of 1914. A reference to her work is kept online at the Experiment Station of the Kansas State Agricultural College, Manhattan, Kansas, Bulletin No. 3, 1888.

Jane Webb Loudon (1807 – 1858 CE) was a British botanist who wrote popular level books on botany, including *The Ladies Flower Garden*, which would eventually be published in four volumes. Copies of her floral prints are still available.

Sarah Plummer Lemmon (1836 – 1923 CE) was a botanist who served as a hospital nurse during the Civil War. She discovered a new genus of plants. She was a member of the California Academy of Sciences and successfully lobbied to make the California poppy (Figure 39) the state flower.

FIGURE 39 field of poppies

Helen Dean King (1869 – 1955 CE) was a botanist who received a PhD from Bryn Mawr College in 1899. She was a member of the National Academy of Sciences who published prolifically on her work including her study of 25 generations of albino rats to examine the effects of inbreeding. Ogilvie has a nice biography of her in her book.

Ida Henrietta Hyde (1857 – 1945 CE) became the first woman to receive a PhD from Heidelberg University (1896). After becoming the first woman to be permitted to do research at Harvard Medical School, she became a professor of physiology at the University of Kansas. She continued her research at the Woods Hole Laboratory during the summers, studying the physiology of marine animals. In addition, she developed the microelectrode which revolutionized neurophysiology (the tiny electrode could directly contact and therefore measure the response of the neurons). In response to her sometimes unenviable position as a pioneering woman, she wrote a darkly humorous piece, *Before Women Were Human* published in the *Journal of the American Association of University Women*, 1938. It described her experiences at Heidelberg. Upon passing her final examination the examining committee would not award the *summa cum laude* that she deserved, instead inventing the *multa cum laude superavit* (praiseworthily excellent in many things).

Clara Eaton Cummings (1853 – 1906 CE) was another early member of the Wellesley Department of Botany. She came to Wellesley as a student in 1876 and kept her connection with the college as an associate professor until her death. She was a scholar of distinguished reputation. Her work in cryptogamic (a plant that produces spores) botany thrust Wellesley into the forefront of research.

Lydia White Shattuck (1822 – 1889 CE) was a botanist who studied at Mt. Holyoke College and then taught there. She became a prominent botanist, known internationally. The school has a nice web site about her. Shattuck Hall at Mt. Holyoke College is named for her.

Mary Agnes Meara Chase (1869 – 1963 CE) was a botanist who became an illustrator at the US Department of Agriculture. During her work at the Bureau, she made great contributions to the study of grasses (agrostology), and her work had important applications for agriculture. She collected over 4,500 specimens from Brazil, Mexico, Puerto Rico, and the United States, and later donated her collections to the Smithsonian and the National Herbarium. She was active well into her senior years going on a field trip to Venezuela when she was in her seventies. One of her publications was the authoritative *Manual of the Grasses of the United States*, 1950.

Elizabeth Knight Britton (1858 – 1934 CE) was a botanist who graduated from what is now Hunter College in New York City. She published 346 professional papers, an impressive amount for any scientist even today! She is credited with being the first person to suggest the establishment of the New York Botanical Gardens. In her later years she participated in the founding of the Wild Flower Preservation Society of America, helping to push conservation measures through the New York legislature. Fifteen species of plants are named for her.

Mary Katherine Layne Brandegee (1844 – 1920 CE) was a botanist who received an MD from the University of California, San Francisco. She served as the California Academy of Sciences curator of botany. She was one of a few pioneer western botanists who collected samples throughout California, Baja, and western Nevada.

Carlotta Joaquina Maury (1874 – 1938 CE) was a paleontologist who received a PhD from Cornell University in 1902. She was the younger sister of the astronomer Antonia Maury (see later in this Chapter). She specialized in Antillean, Venezuelan and Brazilian fossil faunas, places she visited often. She was a fellow of the Geological Society of America, a member of the American Association for the Advancement of Science, the

166

American Geographical Society and a corresponding member of the Brazilian Academy of Sciences.

Alice Middleton Boring (1883 – 1955 CE) was a cytologist, geneticist, and zoologist who received a PhD from Bryn Mawr College in 1910. She spent much of her professional life in China, teaching and publishing. Ogilvie has a nice biography of her in *Women in Science*. Ogilvie and Choquette published her biography in *A Dame Full of Vim and Vigor*, Gordon & Breach, 1999.

Alice Eastwood (1859 – 1953 CE) was a botanist with very little formal training who became one of the best systematic botanists of her time. Several plants are named for her, including *Eastwoodia elegans Brandegee* a member of the sunflower family. The California Academy of Sciences web site has several images of her. Harvard College has a nice web site devoted to her life story. The library there has her papers.

Amalie Dietrich (1821 – 1891 CE) was a German naturalist who filled the void left in botany by Josephine Kablick's death (see Chapter 4). Abandoned by her husband, she was employed as a collector by Caesar Godeffroy who sent her to Australia and New Guinea to gather specimens. Amalie helped introduce Australia's natural wonders to Europe. She spent nearly ten years (1863 – 1872) in the barely settled wilds of northern Queensland, collecting for the Museum Godeffroy in Hamburg, Germany. Her collection of spiders still forms a major reference.[19]

Rosa Smith Eigenmann (1858 – 1947 CE) was an ichthyologist (study of fishes). She worked mainly with her husband Carl Eigenmann. She once wrote,

"In science as everywhere else in the domain of thought woman should be judged by the same standard as her brother. Her work must not simply be well done for a woman."

One always hopes this is the case. Part of the struggle during this time as with other times was exactly for this situation – for women to be accepted not as mere weak echoes of men, but simply as scientists.

Elizabeth Cary Agassiz (1822 – 1907 CE) was a writer on natural history. She married Louis Agassiz (1807 – 1873 CE) and spent her life popularizing his work. He was one of the great scientists of his day and one of the "founding fathers" of the modern American scientific tradition. **Rachel Bodley** (1831 – 1888 CE) was a chemist and botanist who held the first chair of chemistry at the Woman's Medical College in Philadelphia. In 1874 she became dean of the school. **Estrella Eleanor Carothers** (1883 – 1957 CE) was a geneticist and cytologist who received a PhD from the University of Pennsylvania in 1916. Her work focused on the cytological basis of heredity. **Anne Botsford Comstock** (1854 – 1930 CE) was a naturalist who lectured at Cornell University and part of that first group of women inducted into Sigma Xi. Her book the *Handbook of Nature Study* (Figure 40) is still in print.

There was an agricultural depression in the United States in the 1890's. To slow down the exodus of farm workers to the city the state of New York appropriated $8,000 to teach nature study in rural schools, the program administered by Cornell. Several women benefited from this program including Anne Comstock.

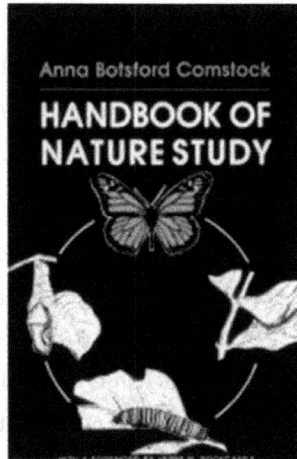

FIGURE 40

Sophia Pereyaslaw (19th century) was a marine biologist whose book was published in 1892, but that is all I can find about her.

FIGURE 41 Chickadee

Florence Merriam Bailey (1863 – 1948 CE) was the first woman to be a fellow of the American Ornithologists' Union (1885). In 1906 the chickadee (Figure 41) was named in her honor: *farus gambeli baileyae*. She wrote several bird guides.

Isabel Cookson (1893 –1973 CE) was a paleobotanist who worked at the University of Melbourne, Australia. She discovered and named *Cooksonia*, one of the most ancient plants known. In 1970 the Paleobotanical section of the Botanical Society of America established a prize in her name called the Isabel Cookson

Paleobotanical Award which is given to recognize an outstanding contribution by a student or recent PhD. She was an intrepid collector of plant fossils. The fossil *Cooksonia* (Figure 42) shown below is courtesy of http://www.xs4all.nl/~steurh/cooksoni.html.

Libbie Henrietta Hyman (1888 – 1969 CE) was a zoologist who received her PhD from the University of Chicago in 1915. She wrote two laboratory manuals and a six-volume work, *The Invertebrates*. The American Museum of Natural History awarded her a Gold Medal Award for Distinguished Achievement in Science, and she also won the Gold Medal in Zoology from the Linnaean Society of London.

FIGURE 42 Cooksonia

Hanna Resvoll-Holmsen, botanist – was a member of a small Norwegian scientific expedition to Svalbard (the Spitsbergen archipelago) in 1907. **Kristine Elisabeth Heuch Bonnevie** (1872 – 1948 CE) was a zoologist and geneticist, for whom the Bonnevie-Ullrich syndrome is named (a congenital syndrome). She studied at the University of Oslo and received a PhD in zoology in 1906. In 1912 she became the first woman to be a professor in Norway when she was appointed professor of zoology. She loved nature "where work and pleasure is so tightly interwoven that you cannot tell where the one ends and the other begins." In 1911 she became the first woman appointed to the Norwegian Academy of Science. In 1908 she published an article that contributed to the establishment of the modern concept of the structure of chromosomes. It took 25 years for her interpretation to be proven right. In a eulogy given at meeting of the Norwegian Academy of Science and Letters, Bjørn Føyn quoted her personal philosophy: "*Age and death follow as natural parts of the life of each subject – in the same way as the plants wither at the end of their flowering period. The individual has done its deed, and life is at an end. But if they have succeeded during their lifetime in arriving at some of the goals of the ethics of Nature, to live according to the best in their characters, then their lives will, without doubt, leave some marks behind among their fellows and relatives.*"

Thus ends a long list of women who specialized in the various subfields of biology and natural history. You can find much more information about each of them on the web. There is one charming addition to this list.

The children's stories by **Beatrix Potter** (1866 – 1943 CE) of England delight us. Yes, the same one who wrote the delightful books for children was a mycologist. She studied the humble

171

fungus. Some of her works are at the Royal Botanical Gardens at Edinburgh.

She showed that algae and fungus belong to the same family, studied spoor germination and the life cycles of fungus. Throughout her life she sketched her surroundings, meticulously recording plants, fossils and animals. From many walks in the woods she amassed a set of detailed watercolors of fungi (Figure 43). The collection (some 270 completed by 1901) is in the Armitt Library, Ambleside, England. She wrote that lichen were really a symbiotic form of algae and fungus. It took many years for her idea to be accepted as correct. She was also quite interested in sheep breeding and became a judge at local sheep fairs. Her scientific work was not accepted during her lifetime, and discouraged she left science behind for her books for children.

FIGURE 43 drawings by Beatrix Potter

Around the time of her death in 1943 many of her notes, including her paper on spores, were burned during the bombing of London in WWII. She kept a private journal which wasn't published until 1966 (because it was written in a code of her own invention). The code was broken by Leslie Linder, an engineer. Once the code was broken it took him seven years to decipher the journal. In the journal she details her attempts to have her theories and drawings noticed, usually to no avail. Perhaps if Beatrix Potter hadn't been so shy, and if the male scientists at the Royal Botanical Gardens hadn't been so dismissive, we would not have her legacy of Peter

Rabbit. Having received posthumously an official apology from the Linnaean Society for its treatment of her (they scoffed at her idea that lichen were a symbiotic form), at a meeting held in her honor in 1997, exactly one hundred years after it had barred her from speaking, she is now beginning to receive the recognition she so richly deserves.

Anthropology/Archeology/Geology

A field new to the sciences is ethnology – cultural and social anthropology. The result of all that exploration in prior centuries inevitably led to the comparative studies of the new cultures. Let us look at five women in this field next.

In 1883 with the publication of *Myths of the Iroquois* **Erminnie Adelle Platt Smith** (1836 – 1886 CE) became the first woman to engage in the field of ethnology. Smith and her family lived in Germany for four years during the 1870s. In Germany, she studied mineralogy and crystallography. She then became an expert on the Six Nations (the Iroquois Nation of New York and Canada), specializing in their languages, and in 1880 she was retained by the Bureau of American Ethnology of the Smithsonian Institution in Washington, DC. Smith was the first woman inducted into the American Academy of Science and was a member of the American Association for the Advancement of Science and the New York Historical Society.

Frances Densmore (1867 – 1957 CE), known as the 'song catcher', was an ethnologist. She loved music and studied at Harvard. She spent her life trying to gather up scraps and artifacts of the old Indian ways, shipping them off to the Smithsonian Institution before an inevitable tide of American progress carried them away.

Alice Cunningham Fletcher (1838 – 1923 CE) was an ethnologist who because a special agent for the US Indian Bureau and eventually a research fellow at the Peabody Museum, Salem, Massachusetts.   She had studied archaeology at the Peabody, starting rather late in her life – in the late 1870's.   In 1881 she arranged to live with and study the Omaha Indians of Nebraska.   In 1889 she moved to the Nez Percé Reservation in Lapwai, Idaho. She brought the scientific rigor of archaeology to the study of ethnology.   She served as vice-president of the American Association for the Advancement of Science (1896), president of the Anthropological Society of Washington (1903), president of the American Folk-Lore Society (1905) and founding member of the American Anthropological Association (1902).   She was the first ethnologist ever to produce a complete description of a Plains Indian ceremony.   The PBS web site has a very nice article about her that includes a photo.[20]

Matilda Coxe Evans Stevenson (1849 – 1915 CE) was the first American ethnologist to turn serious attention to children by publishing the *Religious Life of the Zuni Child*.   She was, perhaps, the first woman ethnologist to work in the American southwest.   In 1885 she founded the Women's Anthropological Society of America.   As a female, she had access to information on Native American women that was inaccessible to male researchers.   Her main interest was in religion and ceremonial traditions, however, leaders of the Smithsonian expeditions encouraged her to deal with issues regarding women and children.   Thus, she is often considered the first American ethnologist to consider women and children as worthy of notice in research.

Ruth Fulton Benedict (1887 – 1948 CE) was an anthropologist (see Figure 44) with a PhD from Columbia University (1922). Her dissertation, "The Concept of the Guardian Spirit in North America", discussed the cultural implications of an individualized religious experience.

FIGURE 44

She became good friends with Margaret Meade. She taught at Columbia University from 1923 to 1948. She wrote of the differences between the cultures around the world and talked about different patterns related to culture and behavior and wrote four books, coming to see primitive culture holistically. In 1934 she published *Patterns of Culture* (Figure 45) which became an American classic and vaulted her to immediate acclaim

It remains one of the most widely read books in the social sciences ever written. She helped to shape anthropology not only for the United States but also the world.

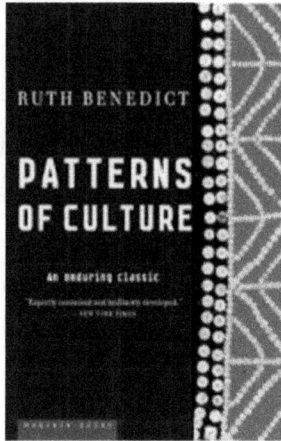

FIGURE 45

A woman I am not discussing is **Margaret Meade** (1901 – 1978 CE) because it is very easy to find information on her work and her life. However, on January 19, 1979, President Jimmy Carter announced that he was awarding the Presidential Medal of Freedom posthumously to Mead. U.N. Ambassador Andrew Young presented the award to Mead's daughter at a special program honoring Mead's contributions, sponsored by the American Museum of Natural History, where she spent many years of her career. The citation reads:

"Margaret Mead was both a student of civilization and an exemplar of it. To a public of millions, she brought the central insight of cultural anthropology: that varying cultural patterns express an underlying human unity. She mastered her discipline, but she also transcended it. Intrepid, independent, plain spoken, fearless, she remains a model for the young and a teacher from whom all may learn."

We drop by to visit an archeologist. Following in the footsteps of the sister of Napoleon (Chapter 4) **Harriet Boyd**

Hawes (1871 – 1945 CE), Smith College class of 1892, was the first archeologist to discover and completely excavate a Minoan town (1901 – 1904), Gournia, east-central Crete (Figure 46).

FIGURE 46 "goddess" figures from Gournia

The Minoan civilization flourished between 3000 and 1450 BCE. The snake goddess shown in Chapter 1 is from Knossos, Crete. Hawes' work is still considered the definitive study. The Smith College archeology department has online photos of her at work. She was the first woman to lecture to the Archaeological Institute of America – ten times in fourteen days in January 1902. In 1992, her daughter, Mary Allsebrook, published *Born to Rebel: the Life of Harriet Boyd Hawes*. The book was edited by Annie Allsebrook, Harriet Boyd Hawes' granddaughter.

177

Geography, with its emphasis on maps, navigation, and exploration, had long been a technical subject. **Ellen Churchill Semple** (1863 – 1932 CE) was a geographer, perhaps the first woman of influence in that field. She studied at Vassar and then at the University of Leipzig. She traveled the remote highlands in Kentucky to study the influence of geographic isolation on the population, the subject of her first professional paper. This work was published in 1901 in the article *The Anglo-Saxons of the Kentucky Mountains, a Study in Anthropogeography* in the *Geographical Journal*. She taught at Clark University in Massachusetts. In 1921, she was elected President of the Association of American Geographers.

Geology also had a long history especially because of the interest in mining ores and precious metals. Modern geology probably started with the presentation of a paper before the Royal Society of Edinburgh in 1788. It was titled *Theory of the Earth; or an Investigation of the Laws observable in the Composition, Dissolution, and Restoration of Land upon the Globe* by James Hutton MD.[21]

Florence Bascom (1862 – 1945 CE) was a geologist. She studied mineral crystals and also studied metamorphic rocks. She earned two bachelor's degrees, a master's degree, and a PhD in geology. She was the second woman ever to earn a PhD in geology in the United States and the first at Johns Hopkins University in 1893. While at Johns Hopkins she had to sit behind a screen in the classroom so as not to distract the men in the class. She then taught geology at Bryn Mawr College in 1895. She was the first woman elected a fellow of the Geological Society of American (1894) and the first to receive an appointment to the US Geological Survey (1896).

178

Chemistry

Moving to the 'hard' sciences, we find an increasing emphasis on quantization and mathematical rigor. Science had become a fully endowed profession as opposed to an avocation often dependent upon patronage for its support.

The history of modern chemistry probably began with the distinction of chemistry from alchemy by Robert Boyle in his work *The Sceptical Chymist* in 1661. By the 19th century we find that chemistry was already maturing into subspecialties. The two women's colleges, Mt. Holyoke and Bryn Mawr, had strong research groups in chemistry.

Mary Engle Pennington (1872 – 1952 CE) was a food chemist who received her PhD in 1895 from the University of Pennsylvania. In 1898 she began teaching at Women's Medical College, specializing in bacteriology. She was the first women to become a member of the American Society of Refrigerating Engineers. She worked for the US Department of Agriculture for several years before establishing her own consulting office in New York City and was instrumental in implementing the new (1908) Federal Food and Drug Act. She not only carried out careful scientific studies on the relationship between handling conditions and bacterial levels in milk and milk products, she also used her work to persuade farmers to agree to new procedures that would help keep milk and milk products safe for the consumer. Because of her thorough and careful work, she was able to define procedures for every step of the process of taking chickens from the slaughterhouse to the consumer. This was one of her greatest contributions. In 2002 she was inducted into the National Women's Hall of Fame.

Ellen Swallow Richards (1842 – 1911 CE) was a chemist and home economist. She studied at Vassar, and was then accepted as a special student at the Massachusetts Institute of Technology (MIT) becoming, I think, the first woman to be accepted at a scientific institution in the United States. She subsequently taught at MIT. From 1887 to 1897 Richards served as official water analyst for the State Board of Health. You can also find an article on her in the online *Encyclopedia Britannica*. She helped to found what is now known as the Marine Biological Laboratory at Woods Hole, Massachusetts (http://www.mbl.edu/). She is sometimes called the "woman who founded ecology".[22] She was the first woman elected to the American Institute of Mining and Metallurgical Engineers.

Agnes Pockels (1862 – 1935 CE), German, was largely self- taught in chemistry. She was one of the founders of surface chemistry (the surface tension of liquids). In 1881, beginning a forty year career in research, she observed the streaming of currents when salts were put into solution and, by attaching a float to a balance, measured the increase in surface tension. For ten years she studied the properties of surfactants (materials that can reduce the surface tension of water) and surface tension of liquid solutions in her own home. Then when Lord Rayleigh (winner of the Nobel Prize in physics in 1904) began to publish on this subject, she wrote to him (in German) about her work. Rayleigh found it so remarkable that he had it translated into English and sent it for publication to the scientific journal *Nature*.

The Nobel Prize was established by Alfred Nobel, the Swedish chemist and munitions manufacturer who invented dynamite. These are the most prestigious awards in the world. The first prizes – in chemistry, physics, medicine, literature, and peace – were awarded in 1901. An additional prize in economics was established in 1969. By the terms of Nobel's will, there can be no prize for mathematicians/astronomers. Sir Run Run Shaw established the Shaw Prize, I believe in 2004. He allowed astronomy and mathematics. These are often called the Far East Nobel Prizes. In 2007 there were three Shaw prizes: one in astronomy, one in mathematics, and one in life sciences and medicine.

Katherine Burr Blodgett (1898 – 1979 CE) earned her BS in chemistry from Bryn Mawr (1917) and her MS from the University of Chicago (1918). In 1926 she became the first woman to receive a PhD in physics from Cambridge University. She spent her research life at the General Electric Company. She started there in 1918 benefiting from the labor shortage due to World War I. She worked in a group developing monomolecular films. Her work with Irving Langmuir was groundbreaking. This type of "invisible glass" is now standard fare in cameras and optical equipment. For her work she received the American Association of University Women's Achievement Award in 1945 and the Garvan Medal from the American Chemical Society in 1951.

In the northeastern US the chemistry research group at Mt. Holyoke produced many chemists of note. **Emma Perry Carr** (1880 – 1972 CE) received her BS and PhD from the University of Chicago and spent her career at Mt. Holyoke. In 1937 she received the first Garvan Medal ever awarded for her work on the electronic spectra of aliphatic hydrocarbons in the far ultraviolet. Others from Mt. Holyoke are mentioned in Chapter 6.

Onward to other technical fields…

Psychology

Now we move to a new field – psychology, just developing during the 19th century and coming to full flower in the 20th century. Women contributed here just as they contributed to other sciences. I have list of seven.

In 1893 the psychologist **Millicent Washburn Shinn** (1858 – 1940 CE) published *Notes on the Development of a Child*. In 1898 she became the first woman, and the eleventh person, to receive a PhD from the University of California, Berkeley. Her book *The biography of a baby* is still used as a college text.

Margaret Floy Washburn (1871 – 1939 CE) was an American psychologist. Finally settling at Vassar after a long career in teaching, she published prolifically. She wanted to bring together philosophy and science in the new field of experimental psychology. She authored over sixty professional papers, and her major book was *The Animal Mind* published in 1908. She was the second woman elected to the National Academy of Sciences (1931).

June Etta Downey (1875 – 1932 CE) was a psychologist who got her PhD from the University of Chicago in 1907. She published seven books and about seventy professional papers. The University of Wyoming where she taught named a building after her. A gifted and often ingenious experimenter, she followed her principal interest in the psychology of aesthetics into many areas of the arts and the mental processes associated with them. Work in muscle reading, handwriting, handedness, color perception, and such topics led to deeper investigations into personality and creativity. Her research on handwriting and other motor functions led to the development of the Downey Individual Will-Temperament Test, an early personality inventory. It was one of

182

the first tests to evaluate character traits separately from intellectual capacity.

Christine Ladd-Franklin (1847 – 1930 CE) was a psychologist who received her PhD from Johns Hopkins University. Because the school was not awarding PhD's to women in 1882, she left even though her dissertation was written; they did not award her the degree until 1926, forty-four years after her dissertation. Her main work was in the field of color vision. She was a psychologist, a logician, a mathematician, and at times an aspiring physicist and astronomer. She was known for her bold expression of her ideas and theories in an academic environment that was often less than welcoming, and she supported and encouraged women to pursue an education.

Lillien Jane Martin (1851 – 1943 CE) was a psychologist who received her PhD from the University of Göttingen (Germany) in 1898. She then joined the staff at Stanford where she progressed through the academic ranks eventually becoming chair of the psychology department, the first woman to head a department at Stanford. Her determination eventually rewarded her with an honorary PhD in 1913 from the University of Bonn. Lillien's accomplishments and enthusiastic eagerness to share knowledge changed the way applied psychology is viewed in the areas of gerontology and mental hygiene for children.[23]

Karen Horney (1885 – 1952 CE) was a psychologist, a giant of psychoanalysis who specialized in extensions to Freudian analysis especially with respect to women showing they were not Freud's idea of womenhood. There is an excellent biography of her life: *A Mind of Her Own*, by Swan Quinn.[24]

Mary Whiton Calkins (1863 – 1930 CE) was a psychologist and philosopher who taught at Wellesley College.

Although she completed the requirements for a PhD, Harvard would not grant her the degree because she was a woman. A pioneer in her field, she was among the first generation of women to enter psychology. She fought for access to Harvard's seminars and laboratories. With her tenacious attitude, she moved on and opened one of the first psychological laboratories in the United States at Wellesley College in 1891. She kept her faculty position at Wellesley until her retirement forty years later. Her numerous contributions to society included the invention of the paired-associate technique for studying memory, groundbreaking research on dreams, and the development of a form of self-psychology. Furthermore, she became the first woman to be president of both the American Psychological Association and the American Philosophical Association.

Engineering

Women quickly became leaders in the emerging field of psychology. What about other fields? Engineering, for example, had long been a bastion of men. But with access to colleges, a few brave women soon joined the ranks of the engineers. It was a time when talented and energetic engineers could make enormous personal contributions, and when human welfare and technological progress were often assumed to walk hand in hand. For a while, James Watt (the steam engine) and his successors were the superstars of their age.

In 1876 **Elizabeth Bragg** (1860 – 1899 CE) received the first civil engineering degree awarded to an American woman (from the University of California at Berkeley). **Bertha Lamme** (1869 – 1943 CE) received the second such degree in 1893 from Ohio State University. She worked at Westinghouse, and in 1973

184

they established the Westinghouse/Bertha Lamme Scholarships in honor of the first woman engineer employed at Westinghouse.

Emily Roebling (1843 – 1903 CE) was married to the engineer in charge of building the Brooklyn Bridge (Figure 47). When he became ill, she took over the job of managing this outstanding engineering feat. She became the first woman to address the American Society of Civil Engineers speaking on behalf of her stricken husband. In 1881 she and the trustees were the first to walk across the newly opened span of the bridge. Her name is inscribed on a piling at the bridge.

FIGURE 47 The Brooklyn Bridge

In 1875 **Hannah Clapp** (1824 – 1908 CE) won the contract to build the iron fence surrounding the Nevada state capitol. The web site of the Nevada State Library has a nice article about her.

Multi-talented, she also was the first professor of history and the English languages at the University of Nevada in Reno.

Julia Brainerd Hall (1859 – 1925 CE) was a chemical engineer. Iron and steel are relatively dense, heavy metals. Aluminum is about one-third as dense as iron and is the most common metal in the Earth's crust. The main difficulty in using the lighter aluminum instead of iron and steel was extracting aluminum from its ores. In 1856 the price of aluminum was $90 per pound (1856 dollars) making it practically a precious and exotic metal. Julia worked with her brother Charles on a process for a commercially viable route to producing aluminum, known as the Hall process. By using this process the cost of producing aluminum was lowered to reasonable limits (about $2 per pound, eventually reaching 30¢ a pound), and she and her brother joined with Paul Héroult to found a company now known as Alcoa (Aluminum Company of America).

Kate Gleason (1865 – 1933 CE) was a mechanical engineer well known for her original design of worm gears. She did not have any thorough training in engineering although she did attend Cornell University, as a 'special student' in 1884 to study mechanical arts. She also attended the Sibley College of Engraving and the Mechanics Institute (now the Rochester Institute of Technology) part time while she was at Cornell. She began her career at her father's machine-tool factory where she propelled it into the leading U.S. producer of gear-cutting machinery prior to World War I. During World War I the president of the First National Bank of Rochester resigned to join the military. So from 1917 to 1919 she served as its president, becoming the first woman to be the president of a national bank. During that time she began to promote the large-scale development of low-cost housing and set to work on projects related to this. In 1918, because of her excellent reputation in the housing construction field and machine-tool business, she became the first women elected to the American Society of Mechanical Engineers. She also served as the society's representative to the World Power Conference in Germany in 1930.

Florence Caldwell Jones (1868 – 1937 CE) graduated from the Colorado School of Mines in 1898 with a degree in Civil Engineering. She was first woman in Colorado to receive such a degree and perhaps the third in the United States. Born in Paris, Kentucky, **Margaret Ingles** (1892 – 1971 CE) was the second woman engineering graduate in the United States and the first woman to receive the professional degree of mechanical engineer in 1916 from the University of Kentucky. In 1920 she became the first women to earn a Master's degree in mechanical engineering. She became a pioneer in the development of air conditioning while working for the Carrier Corporation. As a side light, in the year of her birth, 1892, *The Bee*, the newspaper from Earlington, Kentucky wrote a short editorial announcing that there are forty-eight men and one woman in the graduating class of pharmacy at Northwestern University, and the woman, **Viola Griswold**, took the first prize.

The first woman to gain prominence as a professional architect (1881) was **Louise Bethune** (1856 – 1913 CE). She designed many buildings including one of the country's first structures with a steel frame and poured concrete slabs. In 1886 she became the first woman to join the American Institute of Architects.

Mathematics/Astronomy/Physics

And now onto the astronomers, mathematicians, and physicists … There are a number of these as well. Astronomy and mathematics have the longest intact history of the sciences. The American Astronomical Society, the professional society for American astronomers, was formed in 1899. Fifteen percent of the founding members were women! That percentage held firm through most of the twentieth century.

Stepping back just a bit to the beginning of the 19th century to catch an important figure, we find **Augusta Ada Bryon, Countess of Lovelace** (1815 – 1852 CE) an English amateur mathematician. She was the daughter of Lord Byron the poet. She sponsored and supported Sir Charles Babbage (1791 – 1871 CE) who designed the "difference engine", one of the first mechanical computers, although he did not actually build a working model. She developed the first subroutine for this model. This subroutine was finally compiled and run in the mid-20th century. She wrote that *"The Analytical Engine has no pretensions whatever to originate anything. It can do whatever we know how to order it to perform. It can follow analysis; but it has no power to anticipating any analytical relationships or truths."*[25] This is the earliest version of the slogan: garbage in – garbage out; *i.e.*, what humans put into computers determines the value of what computers produce. The United States Department of Defense named its official computer language ADA in her honor. ADA is a Pascal derived computer language that is large and complex, aimed at embedded applications.

I must mention **Marie Skoldowska Curie** (1867 – 1934 CE) who was a physicist and chemist. So much is known about her that I won't discuss her here. She won the Nobel Prize for physics, and then again for chemistry – a different field. She remains to this day the only woman or man to do this. She was the first woman in France to achieve full professorial rank. Not only that, her daughter, **Irene Joliot Curie** also won the Nobel Prize in chemistry (1935).

Tatyana Afanassjewa (1876 – 1964 CE) was a physicist who was active in the field of thermodynamics (the study of temperature and movement of energy). She married Paul Ehrenfest, another physicist with whom she worked. Together they published several papers still referenced today. For critical studies on the foundations of thermodynamics one is usually referred to their research papers.

Hertha Marks Aryton (1854 – 1923 CE) was an early 20th century British physicist who worked in electricity and wrote what became a standard textbook: *The Electric Arc*. In 1904 she was the first woman to read a paper before the Royal Society.[26] It was entitled "The Origin and Growth of Ripple Marks" meaning ripple marks formed in sand. She received the Hughes Medal for her original research into the electric arc and the sand ripples. She also invented and patented an instrument for dividing a line into any number of equal parts.

Marcia Keith (1859 – 1950 CE) became the first full-time instructor in physics at Mount Holyoke College. In 1899 she helped to establish the American Physical Society, the professional association of American physicists.

Margaret Eliza Maltby (1860 – 1944 CE) was a physicist who received her PhD from the University of Göttingen in 1895. She returned to the United States to teach at several colleges including Barnard College. The Committee on Women in Physics (CWP) has a nice web entry about her life and contributions as does the library at the University of California, Los Angeles. In 1900 she became a Fellow of the American Physical Society. In

1926 American Association of University Women established the Margaret E. Maltby Fellowship.

Harriet Brooks (1876 – 1933 CE) was a physicist. She was early researcher in the field of radioactivity. She graduated from McGill University in 1898 and was invited by Ernest Rutherford (a renown physicist who later won the Nobel Prize for his work in the chemistry of radioactive substances) asked her to join his team at McGill where she published a paper jointly with him titled "The New Gas from Radium". In the spring of 1901 she received a graduate fellowship to Bryn Mawr College. She spent some time working in England at the Cavendish Laboratory where she made the first observations of the half-life of thorium "emanation" as radioactivity was then called. She later worked with Marie Curie. She abruptly stopped research after her marriage and retired.

Maria Mitchell (1819 – 1889 CE)[27] was one of the foremost 19th century American scientists, male or female. When asked to name a prominent American scientist, anyone of the 19th century general public in the United States would name Dr. Mitchell. Trained by her father in astronomy she went on to receive many honorary degrees.[28] She discovered a comet receiving gold medals from the King of Denmark, from Switzerland, and the Republic of San Moreno for her discovery. She was the first woman to have a comet named for her. At Vassar she trained generations of young women. Her home has become a Mecca for young students in astronomy. The Maria Mitchell Observatory was founded in her name and still exists on Nantucket next to her home.

- 1848 appointed the first woman to the Academy of Arts and Sciences
- 1853 awarded the first American advanced degree to a woman (Indiana Hanover College)
- 1859 awarded the Medal of Merit from Switzerland and the Republic of San Moreno
- 1865 appointed the first director of Vassar College Observatory
- 1869 appointed to the American Philosophical Society
- 1875 elected president of the American Society for the Advancement of Women

And my personal favorite – in 1893 shortly after her death she was honored by having her name inscribed along with those of the greatest thinkers of the world (people like Einstein and Thales[29]) atop the stone sidings at the Boston College Library! She had a wonderful sense of what science is:

"We especially need imagination in science. It is not all mathematics nor all logic, but is somewhat beauty and poetry."

"The eye that directs a needle in the delicate meshes of embroidery will equally well bisect a star with the spider web of the micrometer."

"Until women throw off reverence for authority they will not develop. When they do this, they will come to truth through their own investigations, when doubts lead them to discovery, the truth they get will be theirs, and their mind will go on and on unfettered."

"In my younger days when I was pained by half educated, loose and inaccurate ways which we all had, I used to say, 'How much women need exact science'. But since I have known some workers

in science who were not always true to the teaching of nature, who have loved self more than science, I have said, 'How much science needs women'."

Maria Mitchell was an inspiration to her students, and a legend at Vassar. Her insights were wonderful. Sewing and measuring stellar positions – both require the same steady hand and eye. She saw, as all good scientists do, that imagination forms an integral part of all scientific endeavors.

She wrote of the beauty and poetry of science. There is a wonderful poem that, for me, describes perfectly what a scientist is. It is by Nina Cassian, born in 1924, Galati, Rumania, and quoted by permission of the author.

I've stitched my dress with continents,
bound the equator round my waist.
I waltz to a steady rhythm, bending slightly.
I can't stop my arms
plunging into galaxies,
gloved to elbows in adhesive gold;
I carry on my arms a star's vaccine
With such greedy sight   No one fears me
my eyelids flutter in the breeze   Except Error
like a strange enthusiastic plant   Who is everywhere.

Mary Watson Whitney (1847 – 1921 CE) was an astronomer invited to attend Professor Peirce's lectures on quaternions (a form of complex mathematics discovered by Hamilton (1805 – 1865 CE)[30]) at Harvard. She had to wait outside the college for him to escort her to the classroom because Harvard was not open to women. She then studied celestial mechanics[31] at the University of Zurich (1873 – 1876). In 1896 her observations of the variable star Nova Aurigae paved the way for the study of other variable stars. In 1899 she became a founding member of the

American Astronomical Society. She assisted Maria Mitchell at Vassar, succeeding her as director of the Vassar Observatory.

Despite the rules against women attending lectures, she got in, you notice. And things improved for women in astronomy. Around the turn of the 20th century the Director of Harvard College Observatory, a man named Pickering, started a group of "computers" at the Observatory. These "computers" were usually women, sometimes men, hired to do the exacting computations required in positional astronomy. They were hired to work not to think. The women were paid less (25 cents an hour) than their male counterparts and had to work in the basement to avoid distracting the men, but they were allowed to work in astronomy. Many of the leading women in astronomy in the United States came from this group, including the following eight women.

Williamina Fleming (1857 – 1911 CE) was an astronomer who became a staff member of this Harvard College Observatory group sponsored by Pickering. She had been Pickering's housekeeper, and he, annoyed at a male colleague, boasted that his housekeeper could do better. So in 1881, Pickering hired Fleming to do clerical work at the observatory. While there, she devised and helped implement a system of assigning stars a letter according to how much hydrogen could be observed. Stars classified as *A* had the most hydrogen, *B* the next most, and so on. Fleming contributed to the massive cataloguing of stars that would be published as the *Henry Draper Catalogue*. In nine years, she catalogued more than 10,000 stars. During her work, she discovered 59 gaseous nebulae, over 310 variable stars, and 10 novae. In 1907, she published a list of 222 variable stars she had discovered. In 1888, she discovered the Horsehead nebula on Harvard plate B2312. William Pickering, who took the photograph, speculated that the spot was dark obscuring matter.

All subsequent articles and books seem to deny Fleming and W. H. Pickering credit, because the compiler of the first Index Catalogue, J. L. E. Dreyer, eliminated Mrs. Fleming's name from the list of objects then discovered by Harvard, attributing them all instead merely to 'Pickering'. But, by the release of the second Index Catalogue by Dreyer in 1908, Mrs. Fleming and others at Harvard were famous enough to receive proper credit. She was placed in charge of dozens of women hired to do mathematical classifications and edited the observatory's publications. In 1899, she was given the title of Curator of Astronomical Photographs. In 1906, she was made an honorary member of the Royal Astronomical Society of London, the first American woman to be so elected. Soon after, she was appointed honorary fellow in astronomy of Wellesley College. She published *A Photographic Study of Variable Stars* (1907) and *Spectra and Photographic Magnitudes of Stars in Standard Regions* (1911).

Antonia Caetana Maury (1866 – 1952 CE) was an astronomer trained at Vassar. She joined the staff at Harvard, only to leave when Pickering denied her the credit she deserved when she discovered (1897) the criterion that allowed astronomers to distinguish between giant and dwarf stars. She became a visiting teacher and lecturer in various cities and colleges. She eventually returned to Harvard, and after Pickering died found some degree of freedom in her research. She never spoke a derogatory word about Pickering and apparently suffered his unjust treatment in silence. **Florence Cushman** (1860 – 1940 CE) was another astronomer and member of that famous group at Harvard College Observatory. **Anna Winlock** (1857 – 1904 CE) was an astronomer on the Observatory staff as were **Eva F. Leland** and **Mabel C. Stephens**. **Henrietta Swan Leavitt** (1868 – 1921 CE) was hearing impaired but this did not deter her from her studies. She volunteered at the

Observatory in 1893 at the age of 25 where she worked on the precise measurements of the brightness of stars. Her most important work was the discovery of the period-luminosity relation of Cepheid variable stars. "It is worthy of notice", she observed, that "the brighter variables have the longer periods." Cepheids are stars that fade and brighten in a regular fashion. This regularity is one of the fundamental building blocks in the yardstick used by astronomers to measure distances in the universe. The distance ladder that reaches from the Moon to the nearby stars to the farthest galaxy include as a vital rung Cepheid stars. Interestedly, in 1925 after her death the Swedish mathematician Gösta Mittag-Leffler wrote to her saying he was seriously inclined to nominate her for the Nobel Prize. Unfortunately Nobel prizes are not awarded posthumously. Who knows what might have happened had she lived longer. Her work was that important.

Annie Jump Cannon (1863 – 1941 CE) joined the Pickering group at Harvard College Observatory and worked there until she retired. An annual award in her name is given by the American Astronomical Society to a selected young astronomer early in her career. Annie was born in Dover, Delaware, studied at Wellesley, became an expert in photography, and traveled through Europe taking pictures. In 1896 she participated in the first x-ray experiments in this country. She assigned over a quarter of a million stars to their place in the great spectral catalog: the *Henry Draper Catalogue* of stars. Her classification is still used today. She became curator at Harvard College Observatory in 1911 and received a permanent position in 1938. She was the first woman to receive a doctor of astronomy degree from Grönnigen (1921) (Germany) and received an honorary degree from Oxford in 1925 (and only then would Harvard consider granting her permanent status). In 1923 she was voted one of the twelve greatest living

American women. In 1931 she received the Draper Award from the National Academy of Sciences. When she was awarded the Medal, the presenter, Harlow Shapley said

"The benign presence of the Brick Building, noted collector of degrees and medals, author of nine immortal volumes, and several thousand oatmeal cookies, Virginia reeler, bridge player, and, especially, the recipient of the Draper Medal of The National Academy of Sciences – the first medal ever bestowed on a woman by that honorable body of fossils and one of the highest honors attainable by astronomers of any sex, race, religion, or political preference."

It was said of her (in echo of Newton) that Miss Cannon has mapped the field, and if the view we see is distant and splendid, it is because we stand on her shoulders.

She herself said that

"Classifying the stars has helped materially in all studies of the structure of the universe. No greater problem is presented to the human mind."

Not all of the late nineteenth century astronomers were part of Pickering's original group. **Mary E. Boyd** taught at Smith College, **Susan Cunningham** (1842 – 1921 CE) studied at Vassar, worked with Maria Mitchell, and then taught at Swarthmore College where she planned and equipped the first observatory there. In 1891 she was elected to the New York Mathematical Society, one of the first six women to join. **Hannah Mace** (1870 – 1958 CE) was a "computer" at the US Nautical Almanac Office – the place that prepares *The Astronomical Almanac* and *The Nautical Almanac* for the United States. Founded in 1842 by the US Congress to provide crucial calculations for celestial navigation, the Office has had several directors.[32] Ms Mace graduated from Vassar in 1890 and began her career at the Nautical Almanac Office in 1894.

Dorothea Klumpke Roberts (1861 – 1942 CE) was an internationally known astronomer. A wonderful article was written about her life by Robert Aitken in *Publications of the Astronomical Society of the Pacific*.[33] The Klumpkes wanted the best training for their children, both boys and girls. So Mrs. Klumpke took the children to Europe and put them in schools in France and Germany. Dorothea in due time was admitted to the University of Paris where she received the Docteur ès Sciences, the first women to do so. She then went to work at Paris Observatory where she supervised the production of the French part of the great *Carte de Ciel* – an astronomical catalog of the sky.

In 1893 the Paris Academy of Sciences elected her Officier d'Académie. She married an astronomer, Dr. Isaac Roberts, moved to England, and became his assistant. In 1934 she received the Chevalier de la Légion d'Honneur (Figure 48) and received the Cross of the Legion at the hands of the President of the Republic of France. She returned to San Francisco to live where she continued her interest in astronomy, helping young researchers with their careers.

FIGURE 48

Margaret Lindsay Murray, Lady Huggins (1848 – 1915 CE) was truly a Renaissance woman (Figure 49), being accomplished in music, painting, writing, an expert on antique furniture, and astronomy. Her interest in astronomy began early when her grandfather taught her the constellations. She began to construct homemade instruments and finally arranged to meet William Huggins, a spectroscopist, through the telescope maker

197

Howard Grubb of Dublin. She married Huggins in 1875 and they became close collaborators. Together they produced some of the earliest spectra of astronomical objects, most notably the Orion nebula.[34] She was good friends with Sarah Whiting, the director of the Wellesley College Observatory and so, near the end of her life, gave a great many personal items to Wellesley College. These include many antiques, books, a collection of her drawings of British churches, and a collection of small astronomical instruments including the spectroscopes. Included among the books were the manuscript notebooks of William Huggins.

FIGURE 49 Lady Huggins

Caterina Scarpellini (1808 – 1873 CE) was an astronomer, statistician, and meteorologist. She discovered a comet in 1854 and in 1872 Italy struck a gold medal honoring her work in statistics. She organized the Meteorological Ozonometric station in Rome.

Janet Taylor (19th century CE) was a British astronomer who in 1855 built a beautiful sextant for the Prince of Wales now on display at the National Maritime Museum (Figure 50). Note the three feathers for the Prince of Wales in the center. She was one of the few female instrument makers in London.

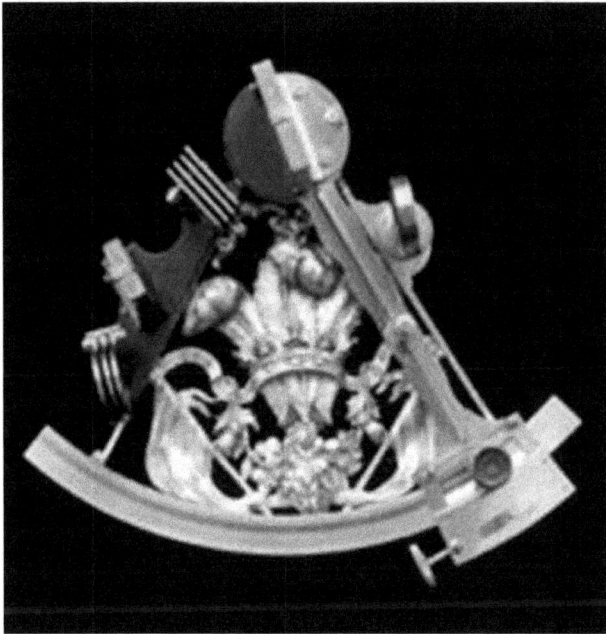

FIGURE 50 sextant

At least two women contributed to astronomy while living in India during the British occupation. **Elizabeth Isis Pogson Kent** (19th century CE) was a British astronomer who worked at the Madras Observatory[35] in India from 1873 to 1896. **Mary Orr Evershed** (1867 – 1949 CE) was an astronomer who spent part of her research career in India at the Kodaikanal Observatory.[36] **Grace A. Cook** (19th century CE) was an astronomer about whom I know very little. I have only her name.

Mary A. Albertson (died 1914 CE) was a botanist and astronomer. She became one of the curators of the Maria Mitchell Observatory in Nantucket. **Anne Sewell Young** (1871 – 1961 CE) was an astronomer who received a PhD from Columbia University (1906) and taught at Mt. Holyoke College. **Flammetta Worthington Wilson** (1864 – 1920 CE) was a British astronomer who studied meteors (observing over 10,000), the aurora borealis, the zodiacal light, and comets. **Sarah Frances Whiting** (1846 – 1927 CE) was yet another astronomer and physicist and an inspiring teacher to a generation of women. While at MIT she met a physics professor named Edward Pickering, the same Pickering mentioned above. He left MIT to become the director of the Harvard College Observatory in 1877. He invited Whiting to observe some of the new techniques which were being applied to astronomy, in particular, spectroscopy. Inspired by this, she introduced the teaching of astronomy at Wellesley in 1880. Teaching the class brought home to her what a necessary and invaluable tool an observatory was. For two decades, she taught astronomy with only a celestial globe and a 4-inch portable telescope. She set herself the task of having an observatory installed at Wellesley. The funding needs for the project were met by a Wellesley trustee and friend, Mrs. John C. Whitin. Whiting drew up the plans for the facility which housed a 12-inch refracting telescope with spectroscope and photometer attachment, and a transit instrument. Once again, as with the undergraduate physics lab, she did most of the work herself. The Whitin Observatory – named after its benefactress and built of white marble to pay homage to the names of the two women who made it happen – was officially opened and established in 1900. Whiting's lifelong commitment to teaching women physics and astronomy, her enthusiasm for the experimental method, and her establishment of the first physics laboratory for women in the United States helped generations of women practice and understand science. She stands as one of the pioneers of science education for women. She taught Annie Jump Cannon (see this chapter).

Amalie Emmy Noether (1882 – 1935 CE) was a mathematician who received a PhD from the University of Erlangen in 1907. Her work was original and creative and formed the basics for much of modern physics. Noether's Theorem (work done in 1915) is named for her. It is one of the basic structures of theoretical physics. She is known for her work in non-associative algebras. Denied recognition for her work most of her life, finally in 1932 she received the Alfred Ackermann-Teubner Memorial Prize for the Advancement of Mathematical Sciences. When she died her obituary was written by Albert Einstein who said

"In the judgment of the most competent living mathematicians, Fraulein Noether was the most significant creative mathematical genius thus far produced since the higher education of women began..."

Quite a tribute from the man many consider the foremost genius of the 20th century.

Annie Russell Maunder (1868 – 1947 CE) was an astronomer trained in Belfast and Cambridge. She became a "computer" at the Greenwich Royal Observatory and editor of the *Journal of the British Astronomical Association*. Another English astronomer of this era was **Ellen Mary Clerke** (1840 – 1906 CE) who wrote popular level articles. She was the sister of **Agnes Mary Clerke** (1842 – 1907 CE) who was also an astronomer who wrote popular level material on astronomy. Her popular level book soon became a standard. In 1903 she joined the select group of women chosen as an honorary member of the Royal

Astronomical Society. **Elizabeth Brown** (1830 – 1899 CE) was an amateur astronomer who published papers on solar eclipses and traveled on eclipse expeditions to Russia and the West Indies.

Mary Adela Blagg (1858 – 1944 CE) was another English astronomer. In the middle of her life she attended a university extension course in astronomy and was so talented that her tutor suggested she develop a uniform lunar nomenclature. At the time, there were great discrepancies among the maps of the moon. In 1905 a committee was appointed by the International Association of Academies to fix the problem. Mary was appointed by the committee to collate the names of all the lunar formations, and her list was published in 1913. In 1920 she was appointed to the Lunar Commission where she continued to work on the nomenclature, and she finally completed the work in 1935. This list became the standard authority on lunar nomenclature, and in her honor, a lunar crater was named after her. She also worked on variable stars and her work appeared in the professional literature. In 1916 she became the first woman elected to membership in the Royal Astronomical Society.

Charlotte Angas Scott (1858 – 1931 CE) was a mathematician who trained in London (one of the first women to obtain a PhD in mathematics there) and then founded the mathematics department at Bryn Mawr College where she continued to teach. She served as the vice-president of the American Mathematical Society in 1906 – the first woman to do so. (It was seventy years before another woman would serve as vice-president.) She published thirty professional papers in her field of algebraic geometry. She also developed the College Entrance Examination Board.

Sofya Kovalevsky (1850 – 1891 CE) was a Russian mathematician (Figure 51). Tradition says that she learned her early mathematics from the wallpaper on her nursery. Her parents could not afford real wallpaper and so they used paper they discovered in their attic which turned out to be lectures in calculus! She worked with some very good teachers but could not matriculate at a Russian university. In 1874 she received a PhD *in absentia* from the University of Göttingen. Her dissertation work is now known as the Cauchy-Kovalevsky Theorem. She applied for a job at the University of Stockholm. Although she got the job, a well-known playwright wrote an article in the newspaper objecting to her appointment:

"As decidedly as that 2 + 2 = 4, what a monstrosity is a woman who is a professor of mathematics, and how unnecessary, injurious and out of place she is."

Strindberg

FIGURE 51 Sofya Kovalevsky

Foolishness is not limited to a particular century. Yet his article did not keep them from giving her the job and did not keep them from electing her to the Russian Academy of Sciences. Kronecker (the mathematician of the Kronecker Delta function) said of her that

The history of mathematics will speak of her as one of the rarest investigators.

This is a much better epitaph than the one by Strindberg. Ogilvie has a nice biography of her in *Women in Science*. In January 2000, we celebrated her 150th centennial jubilee anniversary (Figure 51). Professionals in math, mechanics, physics, astronomy, history and literature acknowledge her heritage. Her legacy was nine major works: three of pure mathematics, one on celestial mechanics, two on the physics of crystals, and three on rigid body rotational motion. She said that:

It is impossible to be a mathematician without being a poet in soul.

Another important Russian mathematician was **Pelageya Polubarinova-Kochina** (born 1899 CE) who contributed to the theory of complex functions (that pesky *i* that shows up as the square root of −1) and the analytic theory of differential equations.

The first PhD in the United States in mathematics that went to a woman was awarded to **Winifred Haring Edgerton** (1862 − 1951 CE) by Columbia University in 1886. One condition of her admission was that she dust the astronomical instruments. She helped to found Barnard College. She opened the door for other women at Columbia. In fact under her portrait at Columbia is the phrase "she opened the door".[37]

Inventors

Of course, women continued their inventiveness. **Margaret Knight** (1838 – 1914 CE) was such an inventor and sometimes called "the female Edison". We own the flat bottomed paper bag to her inventiveness. Prior to her invention paper bags had been more like envelopes. She went on to invent many other things including a shoe-cutting machine, a window frame and sash, and components for rotary engines and motors, for a total of ninety-two inventions and twenty-two patents. Another inventor of this time was **Mrs. A. H. Manning** of Plainsville, New Jersey, who invented a mower and reaper. **Mrs. Lefebre** in 1859 received the patent for the production of nitrates. **Amanda Theodosia Jones** developed the vacuum canning process to preserve food. Her process became the standard canning method in this country following its patent grant in 1873. She also invented a vacuum process for drying food. In 1914 **Mary Phelps Jacob** patented the elastic brassiere. All modern brassieres are modifications of her original design.

FIGURE 52 wringer washing machine

The first patent for a washing machine – manually operated – was in 1691. In 1782 came the hand operated rotating drum. Washing clothes was a very difficult and physically demanding task. This meant that clothes were washed only rarely.[38] This contributed not only to an odorous population but also to the spread of disease. **Ellen Eglui** invented the type of washing machine that had the roller on top for pressing the water out of the clothes (the clothes wringer, Figure 52). This was a *major* advance in washing machine technology because all of a sudden it took less physical effort to wash clothes leading to a radial improvement in the general hygiene of the population. In 1888 she sold the patent rights because

"You know that I am black and if it was known that a Negro woman patented the invention, white ladies would not buy the wringer, I was afraid to be known because of my colour, in having it introduced into the market, that is the only reason."

Following this thought of mechanical improvements – the 19th century brought with it the sometimes questionable fruits of the industrial revolution. The French chemist Philippe Le Corbeiller (1891 – 1980) commenting on the 19th century change in the need for domestic labor stated *"… mechanical power has changed the status of woman in our society. It has brought it about that the motor which one worker in a plant has to control is operated by pushing a button or twisting a dial, and this can be done by a girl without expenditure of strength; she may well be more useful in that capacity than a big burly man with brawn and perhaps less brains. This means economic independence, the possibility of getting away from the family cell if she wants to and cares to. … It creates in her a completely different psychology – one of self-reliance and independence."*[39]

With increasing self-reliance and a growing belief in their own competence women opened school doors in the United States, even if sometimes a bit forcefully, and some women were finally recognized for their work. It was a long time coming, but it did come. The contributions of women, enhanced by their special creativity, flowed a bit more freely and proved that "we especially need imagination in science".

AS THE YEARS PASS

19th century   Maddalena Canedi-Noe
Maria Vittoria Dosi
Martha Carey Thomas
Elizabeth Blackwell
Harriet Hunt
Marie Zakrzewska
Esther Hawks
Sarah Read Adamson Dolley
Lydia Folger Fowler
Edith Jane Claypole
Elizabeth Garrett Anderson
Mary Walker
Rebecca Lee Crumpler
Mary E. Britton
Lucy Hobbs Taylor
Ida Gray
Mary Putnam Jacobi
Nettie Maria Stevens
Frances Elisabeth Crowell
Alice Bennett
Sarah Stevenson
Mary Harris Thompson
Martha G. Ripley
Lydia Adams Dewitt
Florence Nightingale
Linda Richards
Mary Lua Adelia Davis Treat

Cornelia Clapp

Emily Ray Gregory

Emily Gregory

Graceanna Lewis

Ethel Sargent

Susanna Phelps

Mary Murtfeldt

Jane Webb Loudon

Sarah Plummer Lemmon

Clara Eaton Cummings

Lydia White Shattuck

Elizabeth Knight Britton

Mary Katherine Layne Brandegee

Amalie Dietrich

Rosa Smith Eigenmann

Elizabeth Cary Agassiz

Rachel Bodley

Anne Botsford Comstock

Sophia Pereyaslaw

Florence Merriam Bailey

Beatrix Potter

Erminnie Adelle Platt Smith

Alice Cunningham Fletcher

Matilda Coxe Evans Stevenson

Ellen Churchill Semple

Florence Bascom

Ellen Swallow Richards

Agnes Pockels

Millicent Washburn Shinn

Christine Ladd-Franklin

Mary Whiton Calkins

Elizabeth Bragg

Bertha Lamme

Emily Roebling

Julia Brainerd Hall

Viola Griswold

Hannah Clapp

Louise Bethune

Augusta Ada Bryon, Countess of Lovelace

Harriet Brooks

Maria Mitchell

Mary Watson Whitney

Williamina Fleming

Anna Winlock

Eva F. Leland

Mabel C. Stephens

Henrietta Swan Leavitt

Susan Cunningham

Mary E. Boyd

Margaret Lindsay Murray, Lady Huggins

Caterina Scarpellini

Janet Taylor

Elizabeth Isis Pogson Kent

Mary Orr Evershed

Grace A. Cook

Charlotte Angas Scott

Mary A. Albertson

Flammetta Worthington Wilson

Sarah Frances Whiting

Annie Russell Maunder

Ellen Mary Clerke

Agnes Mary Clerke

Elizabeth Brown
Charlotte Moore Sitterly
Sofya Kovalevsky
Margaret Knight
Mrs. A. H. Manning
Mrs. Lefebre
Amanda Theodosia Jones
Ellen Eglui

Turning the 20th century

Josephine S. Baker
Agnes Mary Claypole
Florence Rena Sabin
Anita Newcomb McGee
Shih Mai-Yu
Martha Tracy
Sister Kenny
Mary Davies Swartz Rose
Gulielma Lister
Jennie Arms Sheldon
Mary Jane Rathbun
Margaret Clay Ferguson
Elizabeth Gifford Peckham
Florence Peebles
Edith Patch
Helen Dean King
Ida Henrietta Hyde
Mary Agnes Meara Chase
Carlotta Joaquina Maury
Alice Middleton Boring
Alice Eastwood

Estrella Eleanor Carothers

Isabel Cookson

Libbie Henrietta Hyman

Hanna Resvoll-Holmsen

Kristine Elisabeth Heuch Bonnevie

Frances Densmore

Ruth Fulton Benedict

Margaret Meade

Harriet Boyd Hawes

Mary Engle Pennington

Katherine Burr Blodgett

Emma Perry Carr

Margaret Floy Washburn

June Etta Downey

Lillien Jane Martin

Karen Horney

Kate Gleason

Florence Caldwell Jones

Margaret Ingles

Marie Skoldowska Curie

Irene Joliot Curie

Tatyana Afanassjewa

Hertha Marks Aryton

Marcia Keith

Margaret Eliza Maltby

Florence Cushman

Antonia Caetana Maury

Annie Jump Cannon

Hannah Mace

Dorothea Klumpke Roberts

Anne Sewell Young

Amalie Emmy Noether
Mary Adela Blagg
Pelageya Polubarinova-Kochina
Winifred Haring Edgerton
Mary Phelps Jacob

---

[1] Maria Mitchell

[2] In 1886 Columbia University awarded the PhD in mathematics to Winifred Edgerton. http://www.agnesscott.edu/lriddle/women/merrill.htm

[3] Francis Willard's address to the Women's National Council

[4] http://www.nlm.nih.gov/hmd/blackwell/index.html

[5] *A Woman doctor's Civil War*, Esther Hill Hawks' Diary, ed. G. Schwartz, University of South Carolina Press, 1986.

[6] Most of who were physicians in the Civil War.

[7] The school did not admit women until 1945.

[8] *Women's Words*, by Mary Biggs, Columbia University Press, 1996, p. 271

[9]
http://www.nlm.nih.gov/changingthefaceofmedicine/physicians/biography_87.html

[10] http://www.nlm.nih.gov/changingthefaceofmedicine/physicians/biography_73.html

[11] Her sister taught music at Berea and taught W.C. Handy who wrote "St Louis Blues".

[12] http://www.defenselink.mil/news/Apr1999/n04301999_9904304.html

[13] Mary Mallon was the first "healthy carrier" of typhoid fever in the United States.

[14] The National Academy of Sciences formed in 1863 by President Lincoln with 50 charter members. There are now over 2000 members whose purpose is to provide scientific advice to the nation.

[15] Women's names were chosen for all the features on the planet Venus.

[16] MIT Press, 1986

[17] The *Who's Who* series started in 1899.

[18] The non-profit Scientific Research Society for scientists and engineers.

[19] Her portrait can be found at
http://www.asap.unimelb.edu.au/bsparcs/exhib/dietrich/dietrich.htm

[20] http://www.pbs.org/weta/thewest/people/d_h/fletcher.htm

[21] Chinese polymath Shen Kua (1031 – 1095 CE) was the first to formulate hypotheses for the process of land formation – geology. Based on his observation of fossils in a geological stratum in a mountain hundreds of miles from the ocean, he deduced that the land was formed by erosion of the mountains and by deposition of silt.

[22] http://jchemed.chem.wisc.edu/JCEWWW/Features/eChemists/Bios is a web site that provides brief biographies of women in chemistry.

[23] http://www.webster.edu/~woolflm/martin.html is a web site with a nice history of her life.

[24] Radcliff Biography Series, Addison Wesley, 1980.

[25] *Mothers of Invention* by Ethlie Ann Vare and Greg Ptacek, William Morrow and Company, 1988, p.226.

[26] Reading a paper before the Society was how a professional paper was accepted for publication.

[27] Her first name is pronounced Mōr-ī´-a.

[28] She went to work in a library so she could study the books on navigation and astronomy, teaching herself German and French to aid in her reading.

[29] Thlaes was the first to try to explain natural phenomena without recourse to mythology – 6th century BCE Greece

[30] The idea for quaternions occurred to him while he was walking along the Royal Canal on his way to a meeting of the Irish Academy, and Hamilton was so pleased with his discovery that he scratched the fundamental formula

of quaternion algebra, $i^2 = j^2 = k^2 = ijk = -1$ into the stone of the Brougham bridge

[31] The study of how stars and planets move – positional astronomy

[32] and the first woman to direct the office did not come until the year 2000, **Sethanne Howard**.

[33] Dec. 1942, **54**, p. 217

[34] Photograph courtesy of Wellesley College archives

[35] The Madras Observatory was founded by the British East India Company in 1786 in Chennai (then Madras). For over a century it was the only astronomical observatory in India that exclusively worked on the stars.

[36] A solar observatory now operated by the Indian Astrophysical Institute

[37] *Women of Science, Righting the Record*, ed. Kass-Simon and P. Farnes, Indiana University Press, 1993.

[38] Unfortunately dirty clothes were a source of disease.

[39] *Engineering in History*, R. S. Kirby, S. Withington, A. B. Darling, and F. G. Kilgour, Dover Publications, 1990, p.497

216

CHAPTER 6
THE 20th CENTURY

Women hold up half the sky

The 20th century saw one of the greatest events in human history. Humans left their home on Earth and began to ride the skies. We traveled to the Moon and back. On June 16, 1963 the Russians put a woman in space – **Valentina Tereshkova** (born 1937 CE) who orbited the Earth 48 times (2 days, 22 hours, 50 minutes) – followed twenty years later by the American astronaut **Sally Ride** (born 1951 CE).[1] Valentina Tereshkova (Figure 53) trained as a cosmonaut with four other women. She was the sixth Russian, the eleventh human being, and the first woman in space. She returned to Earth by ejecting from the space capsule and parachuting to Earth, so it was good that she was a fearless parachutist. She is also an excellent speaker and after her flight traveled the world as a representative of the Russian space program. "Once you've been in space, you appreciate how small and fragile the Earth is" she said. In 2000 she was named Greatest Woman Achiever of the Century by the International Women of the Year Association.

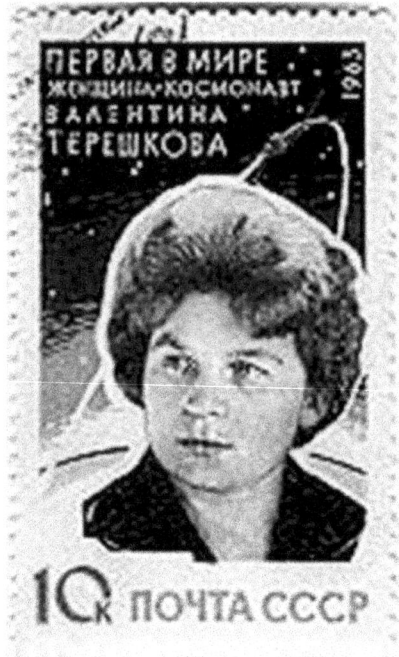

FIGURE 53

The number of women in science and technology from the 20th century is far too large to list in a single book. There are thousands upon thousands of women in science and technology. For example, by the end of the 20th century there were almost 1,000 women in American astronomy – that many women in just one small science (out of a total of about 7,000 American astronomers). This is a truly stunning development. The cycle of women in science has come a full turn since the Middle Ages. Is this true in all countries? I really don't know. Literacy is still the province of the few and lucky. There are places in the world where people starve and die young. So I continue to restrict myself to the Western world and mostly to the United States that I know the best.

For this last chapter I choose a very, *very* small sample – mostly those women about whom I know a little something. It is fairly easy to find information about the technical women in the 20th century. Many have entire books written about them. For example, the book *Women of Science, Righting the Record*, edited by Kass-Simon and Patricia Farnes, 1993, Indiana University Press has an excellent section on women in mathematics who received the PhD from the University of Chicago.

One thing that has changed a lot during the 20th century is the perception of what a woman in science looks like. We were once thought to be old-maidish, stout, and stern – something like this lady on the right (Figure 54). In 1787 a German newspaper described any woman who had scholarly intentions as "one knows in advance that her clothing will be neglected and her hair will be done in antiquarian fashion". But of course we come in all sizes and flavors. Today we are allowed to be as elegant or disheveled as any male scientist.

FIGURE 54 Mrs. Piffleschnick

The 20th century is littered with women of science and technology, and what a joyous picture it is. We can sweep the skies of Earth with women of science and technology. Women even appeared as scientists in fictional literature. Dr. Susan Calvin was the lead engineer of Isaac Asimov's famous science fiction *I, Robot* series. As for the rest, well, a small sample follows.

Engineering

There are still very few women in engineering (only about 10% in 2010) but their numbers are increasing.

What about a movie about an engineer? *Cheaper by the Dozen* is the movie about **Lillian Moller Gilbreth** (1878 – 1972 CE) who was an industrial engineer and perhaps the most famous woman in engineering. In 1931 the Society of Industrial Engineers awarded her the Gilbreth Medal for her pioneer role in economics and modern management.

In 1940 there were just three women and 17,000 men who were members of the American Institute of Electrical Engineering: **Edith Clarke** (1883 – 1959 CE), **Vivien Kellems** (1896 – 1975 CE, rated as the top woman in American industry by the National Association of Manufacturers[2]), and **Mabel MacFerran Rockwell** (1902 – 1981 CE). In 1919 Clarke was the first woman to get a Master's Degree in electrical engineering from the Massachusetts Institute of Technology. She taught at the University of Texas, and when she joined the faculty she became the first woman to be a professor of electrical engineering. She was an authority on electrical power systems and worked on the design and building of several dams in the western United States. The University of Texas, Austin, has a nice obituary of her on their web site.

Elsie Eaves (1898 – 1983 CE) was the first woman admitted to full membership in the American Society of Civil

Engineers (1927). She graduated from the University of Colorado in 1920 with a degree in civil engineering. For more than thirty years the one person who knew more than any other about the statistics of the construction industry in the US was Elsie Eaves. She received the society's highest honor in 1979 – honorary lifetime membership.[3]

Inventors

Women continued their inventiveness. A tiny sample of inventors includes **Bette Nesmith Graham** (1924 – 1980 CE) who invented Liquid Paper (1951). Her company sold to Gillette in 1979 for $47.5 million. Other women were as inventive as she, developing the Melitta coffee filter, washing machines, and the disposable diaper. **Patsy Sherman** (1930 – 2000 CE) invented Scotch Guard, something put on almost all furniture.

Inventions came from non-scientists as well. Many people are familiar with Mrs. Butterworth's maple syrup (Figure 55). What they might not know is that her real name was **Twana Washington** an African-American who published a sophisticated math solution for "smooth polynomials" (making something bumpy into something smooth). She always loved mathematics, but gave into social pressure and became a homemaker. She thought of this new type of smooth polynomial, tradition says, by watching her fresh batch of smooth pancake syrup run off a stack of pancakes.

After her death two electrical engineers stumbled across her journal article. They realized her polynomials were the key to designing filters with a smooth amplitude response. In honor of her pioneering work, these engineers labeled the new devices Butterworth Filters. Some form (Figure 56) of this filter is put in almost every piece of electronics that is built.

221

FIGURE 55

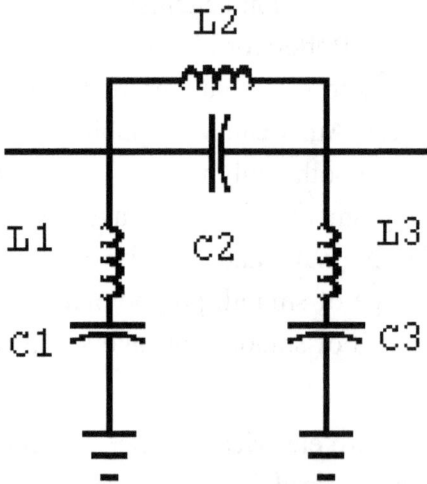

FIGURE 56 27 MHz bandstop Butterworth filter.

FIGURE 57 Hedy Lamarr

Inventors were not limited to mathematician/cooks. The femme fatale (Figure 57) of 1940's Hollywood films **Hedy Lamarr** (1913 – 2000 CE) along with George Anthiel invented frequency hopping to prevent jamming signals sent to torpedoes. They received the patent (#2,292,387) in 1942. The device was a bit too unwieldy for use, but with computerization in the 1960's the invention came into its own. It was renamed spread spectrum technology, and in 1985 the Federal Communications Corporation used it in radios. By the 1990's it was used in cell phones and wi-fi networks. It allows one cell phone to isolate the frequency it wants from the myriad of frequencies in use, so the user can talk without interference. A crucial component of modern technology from an actress!

The Healing Arts

Women were just as inventive in medicine. Although penicillin, the miracle drug, was first identified by Sir Alexander Fleming, he was not the one who pursued its further development. Several of the women who did were: **Gladys Hobby** (1910 – 1993 CE), **Elizabeth McCoy, Dorothy Fennel, Dorothy Hodgkin** (see below), and **Margaret Hutchinson** (1910 – 2000 CE). Dorothy Hodgkin bombarded penicillin with x-rays to deduce how it was put together. Gladys Hobby brewed the first batch of penicillin tested on people. Margaret Hutchinson designed the first commercial plant that made penicillin on a massive scale. Elizabeth McCoy created the strand of penicillin used today.

Dorothy Crowfoot Hodgkin (1910 – 1994 CE) was a chemist. She went to Oxford and then Cambridge to complete her work towards the degree. She stayed at Oxford as an Official Fellow and Tutor in Natural Science at Somerville College (the college named for Mary Somerville, chapter 4). She was part of the team that worked on the initial development of penicillin and vitamin B_{12} and winner of the 1964 Nobel Prize in Chemistry. Many other medical marvels are the result of work by women, including the successful diphtheria antitoxin developed by **Anna Wessel Williams** (1863 – 1954 CE) who then brought a rabies culture to the United States and began vaccine production. **Virginia Apgar** (1909 – 1974 CE) invented the Newborn Scoring System, called the Apgar Score, a world-wide standard used to determine the health of a newborn in the first few minutes of life. She was one of the first women to receive an MD from Columbia University. **Louise Pearce** (1885 – 1959 CE) was one of the main figures in the development of the drug typarsamide. This drug wiped out whole epidemics of African sleeping sickness. She and her colleagues were awarded the Order of the Crown of Belgium (Figure 58).

FIGURE 58 Order of the Crown of Belgium

Helen Brooke Taussig (1898 – 1986 CE) perfected the technique (1945) for saving the lives of "blue babies", babies born with a constriction in a key artery. She was awarded the Medal of Freedom (Figure 59). She also was the first woman to become president of the American Heart Association (1963). She struggled with dyslexia and hearing loss in school, but succeeded despite these handicaps.

FIGURE 59 Medal of Freedom

Rachel F. Brown (1898 – 1980 CE) was the first woman to receive the Pioneer Chemist Award from the American Institute of Chemists (1975). She discovered the vaccine for pneumonia that is still used today. In 1950 along with Elizabeth Hazer she isolated the first antifungal antibiotic effective against fungal diseases. Angela Ferguson (b. 1925) was a relentless researcher who received her MD from Howard University in 1949 and went on to study the disease of sickle-cell anemia. Her work, along with others, has led to the efficient detection and control of this terrible disease. In 1923 microbiologist Gladys Dick (1881 – 1963 CE) and physician George Dick isolated the cause of scarlet fever, and later developed a test for the disease.

The research group at Mt. Holyoke College produced many chemists of note. Mary Lura Sherrill (1888 – 1968 CE) came into her own during WWII. She earned her BA and MA from

Randoph-Macon Women's College in 1909 and her PhD in 1923 from the Unversity of Chicago. She and her students initiated work on anti-malaria drugs for which she received the Garvan Medal from the American Chemical Society. **Lucy W. Pickett** (1904 – 1997) received her PhD (1930) from the University of Illinois. She worked with Dr. Sherrill and Dr. Carr (see Chapter 5) on vacuum ultraviolet techniques for analyzing simple organic compounds. She also received the Garvan Medal.[4]

Dorothy Wrinch (1894 – 1935) was a British mathematician, biologist, and chemist apparently known for her sharp tongue. She advocated a theory of proteins in which amino acids were hooked together in chains subsequently shown to be incorrect although the sequence hypothesis was important. There is a nice article about her in *Women of Science Righting the Record*.[5]

Gertrude Elion (1918 – 1999 CE) was a scientist who studied chemistry at Hunter College in New York City graduating in 1937. She was initially unable to obtain a graduate research position because she was a woman, but she went on to synthesize the leukemia fighting drug 6-mercaptopurine. She also developed drugs to block organ rejections in kidney transplant patients.

In the first half of the 20th century women still had trouble finding jobs commensurate with their abilities. I myself was told in 1963 that it was not a total waste for me to major in physics at college because at least I could teach my children math. What other reason could there be for a woman to study physics? I succeeded despite that disparaging remark. Many women succeeded despite the barriers against them. Things improved dramatically during the second half of the 20th century.

Gertrude Elion did find a job as a lab assistant at the New York Hospital School of Nursing in 1937. She worked as a research chemist at other places finally settling at Burroughs Wellcome Laboratories. There she was first the assistant and then the colleague of George Hitchings, with whom she worked for the next four decades. In 1988 Elion, Hitchings, and Sir James W. Black won the Nobel Prize for Physiology or Medicine for their development of drugs used to treat several major diseases. In 1991 she also received the National Medal of Science and was inducted into the National Women's Hall of Fame.

One of the most influential nurses of the 20th century was **Margaret Sanger** (1879 – 1966 CE). She fought through the deadly tenements of New York City for women's right to birth control. Her papers are available online. She established the principles that a woman's right to control her body is the foundation of her human rights; that every person should be able to decide when or whether to have a child; that every child should be wanted and loved. She forced the reversal of federal and state "Comstock laws"[6] that prohibited publication and distribution of information about sex, sexuality, contraception, and human reproduction. She created access to birth control for low-income, minority, and immigrant women and expanded the American concept of volunteerism and grassroots organizing by setting up a network of volunteer-driven family planning centers across the United States.

Georgia Dwelle (1884 – 1977 CE) was a physician. Daughter of a former slave, she was the first Spelman College alumna to attend medical school, and she established the Dwelle Infirmary in 1920 in Atlanta, Georgia. It was Georgia's first general hospital for African- Americans and its first obstetrical hospital for African-American women. The infirmary, which also featured a pediatric clinic, was Georgia's first venereal disease clinic for African-Americans, and offered Atlanta's first "Mother's Club".

Rosalyn Sussman Yalow (born 1921) was a medical physicist. She received her PhD from the University of Illinois in 1945. As a researcher at the Bronx Veterans Administration Hospital, Yalow and colleague Solomon A. Berson developed a process, called radioimmunoassay (RIA), that made it possible to detect mere traces of biological substances in blood and other fluids. For her work, Yalow was awarded the 1977 Nobel Prize in Physiology or Medicine along with Andrew V. Schally and Roger Guillemin.

Barbara McClintock (1902 – 1992 CE) was a geneticist whose life spanned the 20th century history of genetics. She showed that genes could transpose within chromosomes; that they could move around (the so-called "jumping genes"). This was done through the investigation of maize (corn) genetics through careful hybridization. Her work with genetics came only twenty-one years after the rediscovery of Mendel's principles of heredity, at a time when acceptance of those general principles was not wide-spread. She also traced the evolutionary history of domesticated maize to determine the genetic ancestor of the grass we now call corn. She received the Nobel Prize for medicine in 1983. She also received the National Medal of Science, and her image adorns a US stamp (Figure 60). On her 90th birthday her students and colleagues wrote essays honoring her and her work which were collected into the book *The Dynamic Genome, Barbara McClintock's Ideas in the Century of Genetics* (Plainview: Cold Spring Harbor Laboratory Press, 1992). The picture on the next page (Figure 61) is from Cornell University where she did part of her work.

FIGURE 60

FIGURE 61

World Changing ...

Most people know the name of **Margaret Meade** (1901 – 1978 CE) who published *Coming of Age in Samoa* in 1928. It is still a best-selling anthropology volume. She was perhaps the most famous anthropologist of the 20th century. Indeed, it is through her work that many people learned about anthropology. She wrote over thirty books, lectured widely, and received 28 honorary doctorates (as well as a PhD from Columbia University). She died on the day the *World Almanac* named her one of the world's twenty-five most influential women.

In 1962 **Rachel Carson** (1907 – 1964 CE) wrote *Silent Spring*, the book that brought to the world's attention the long-term effects of uncontrolled pesticides. It was one of the most influential books of its time. She received a master's degree in zoology from Johns Hopkins University. In 1936 she became one of the first women to take and pass the civil service test, and the Bureau of Fisheries hired her as a biologist. She has been called the mother of the modern environmental movement.

The First To ...

The litany of firsts achieved by women started in the last chapter continued in the 20th century.

E. Lucy Braun (1889 – 1971 CE) was the first woman to be president of the Ecological Society of America (1930). She was a professor of botany at the University of Cincinnati and is remembered for her original contributions to ecology. In her honor the E. Lucy Braun Award is given each year by the Ecological Society to an outstanding student.

Marie Gertrude Rand (1886 – 1970 CE) was the first woman chosen as a fellow of the Illuminating Engineering Society, and she received their Gold Medal in 1963. She received her PhD in psychology from Bryn Mawr College in 1911. She studied the way color perception is affected by illumination. For example she and her husband were responsible for the lighting of the Holland Tunnel under the Hudson River (New York City). She held several patents for lighting devices and instruments.

Another first was accomplished by Aerographers Mate First Class (US Navy) **Donna Osif** who was the first woman to forecast the weather in the Antarctic.

Architect **Eleanor Raymond** (1887 – 1973 CE) and chemical engineer **Maria Telkes** (1900 – 1995 CE) built the first solar-heated house. Marie earned a doctorate in physical chemistry and became a pioneer in the use of solar energy. In 1961 Eleanor became a fellow of the American Institute of Architects.

Ida Eva Tacke (1896 – 1979 CE) was a physicist who was the first person to mention the idea of fission in 1934. She discovered element 75 – Rhentium. There are interesting web sites that tell her story.

The physicist **Chien Shiung Wu** (1912 – 1997 CE) was a remarkable lady. Born in China she immigrated to the United States after graduating from Nanking Central University in 1936. She received her PhD from the University of California, Berkeley. She taught at Smith College and Princeton University before taking a position at Columbia University. In 1957, while at Columbia University, she disproved the law of conservation of parity – which had been one of the basic assumptions in physics. She devised the experiment which confirmed a theory proposed by T. D. Lee and C. N. Yang. Lee and Yang received the Nobel Prize for their work but not Chien Shiung Wu. She was, however, given many other honors, including the Comstock Award from the

232

National Academy of Sciences in 1964, the first women to receive this award. She then moved into medical research to study sickle cell anemia. She believed that "even the most sophisticated and seemingly remote basic nuclear physics research has implications beneficial to human welfare." The strands of science extend outward to touch all aspects of human life.

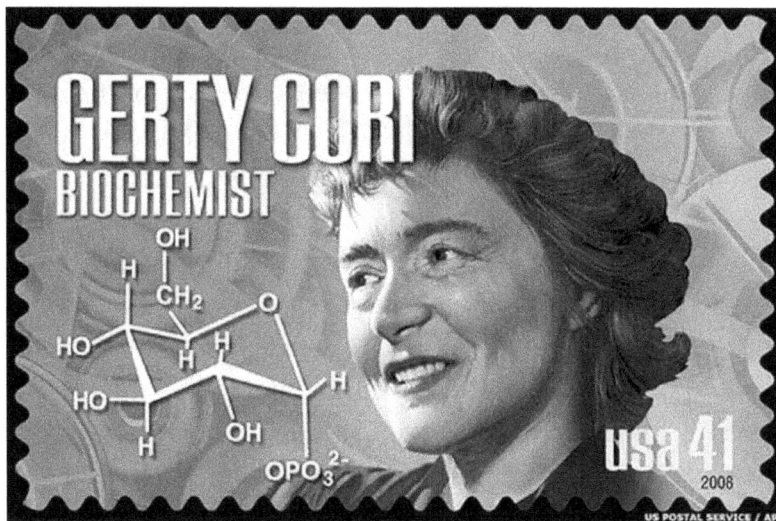

FIGURE 62 Gerty Cori

Biochemist **Gerty Cori** (1896 – 1957), together with her husband won a Nobel Prize in 1947 for discovering how the body processes glycogen. She was the first American woman to win the Prize in Medicine. Born in Prague, she and her husband immigrated to the US. Chemists found a small flaw in the drawing of the "cori ester" molecule, a derivative of glycogen discovered by Gerty, but the US Postal Service issued the stamp (Figure 62) anyway, in March 2008. *"I believe that in art and science are the glories of the human mind. I see no conflict between them,"* is a quote from her essay "Glories of the Human Mind".

World War II

World War II had an interesting effect on women in technical fields. Because many men were gone from their civilian jobs those same jobs were filled by women who performed superbly. "Rosie the Riveter" was real, was successful, and was sent home after the war. When the men returned they replaced the women who were then expected to return happily to a housewife's life. Some women did not take kindly to this and fought to retain access to technical fields. An excellent example of this was **Wanda Alma Marosz** (1923 – 2007 CE). Mrs. Marosz worked for Douglas Aircraft in Los Angeles, designing aircraft wings. *"The men came back from the war, and the company sort of said, 'We're done with you,'"* said her daughter, Kathy Marosz. *"That set her in motion. From that point, she wanted to make sure that women had the same job opportunities as men in math and science careers."* Mrs. Marosz received a bachelor's degree in mathematics from the University of Chicago in 1945 before taking the job with Douglas Aircraft. After losing her job at Douglas, Mrs. Marosz married and went to the University of Southern California, where she obtained a master's degree. She then worked part time as a high school and junior college math instructor. The rest of her life she supported and encouraged girls to study math knowing first-hand the discrimination they felt.

Hero of the Planet

Sylvia A. Earle (born 1935) is a marine biologist who was the first woman to be chief scientist at the National Oceanographic Atmospheric Administration (NOAA), the place responsible for weather forecasts and for monitoring the world's oceans. She holds the women's solo world record dive at 1,000 meters. She still holds the world's record for the deepest solo dive – In 1979 she descended to 1,250 feet, strapped to the outside of a submersible – off the coast of Hawaii. She planted an American flag at the bottom. In 1997 she received the Kilby Award which recognizes individual who make extraordinary contributions to society through science. In 1998 *Time Magazine* named her its first Hero of the Planet (Figure 63).

FIGURE 63 Sylvia Earle

And More Firsts ...

Until **Julia Morgan** (1872 – 1957 CE) applied, no woman had ever been admitted to the exclusive and prestigious L'Ecole des Beaux Arts School of Architecture. Julia succeeded in 1898 and in 1902 received the certification of graduation. She then returned to San Francisco and began work as an architect designing beautiful buildings well matched to their environments. The beautiful and rustic Asilomar Convention Center in Monterey, California exemplifies her vision. By 1919 she had designed over 1,000 buildings in California and was hired by William Randolph Hearst to build the San Simeon estate now known as the Hearst Castle. It took twenty-eight years to build. Her honorary Doctor of Law degree from the University of California in 1929 read in part

> *Distinguished alumna of the University of California; Artist and Engineer; Designer of simple dwellings and stately homes, of great buildings nobly planned to further the centralized activities of her fellow citizens; Architect in whose works harmony and admirable proportions bring pleasure to the eye and peace to the mind.*

Evelyn Boyd Granville (b. 1924) was born in a segregated Washington, DC and sought education as a way out of the prejudiced environment. Her favorite subject was mathematics. With some scholarship aid she entered Smith College in 1941, graduating in 1945 to continue in graduate school at Yale receiving a PhD in 1949. She, **Marjorie Lee Browne** (1914 – 1979 CE), and **Euphemia Loften-Hayes** (1890 – 1980 CE) were the first three African-American women in the United States to receive doctorates in mathematics. After teaching at a few universities she accepted a position at the National Bureau of Standards finally ending up at IBM in 1956. From there she indulged a long time interest in space and worked on a NASA program to track vehicles in space. In 1967 she accepted a university position where she stayed until retirement. Her biography is titled *My Life as a Mathematician*.[7]

236

The first African-American (woman or man) to earn a PhD in geology was **Marguerite Thomas** (1942, Catholic University, 1895 – c. 1991 CE). She was employed as a teacher and then full professor in Miner Teachers College (now part of the University of the District of Columbia) from 1923 to 1955. She also served as an Instructor for the Evening School at Howard University, 1944.

Marie Daly (1921 – 2003 CE) was a biochemist. She was the first African-American woman to earn a PhD in chemistry (Columbia University in 1948). She worked at Howard University, Rockefeller Institute, Columbia University, and Yeshiva University. Her work focused on nucleic acids.

A litany of firsts is nice, but it does not mean the first was the only one to succeed. The first one was followed by a steady and ever increasing stream of women who followed in the footsteps of the ones who opened the doors to science and technology.

Flying

Airplanes were invented in the 20th century and women were right there with the men. **Bessie Coleman** (1893 – 1926 CE) was the first African-American pilot who was a woman (Figure 64). She received her license to fly in 1921 in France. She is another member of the National Women's Hall of Fame. She died in a flying accident in 1926. The first woman to fly solo (1910) in the United States was **Blanche Scott** (1885 – 1970 CE). In 1928 **Amelia Earhart** became the first woman to fly the Atlantic. She repeated the flight solo in 1932, and then in 1935 became the first person to make a solo flight from Hawaii to California.

FIGURE 64 Bessie Coleman

Mrs. Olive Ann Beech (died 1993) was a leading entrepreneur in American aviation. She was awarded the Smithsonian's National Air and Space Museum Trophy in 1993. She became CEO of the Beech Aircraft Company in 1950 upon the death of her husband and continued to make numerous contributions to the aerospace industry. In 1980 Mrs. Beech was presented one of aviation's most distinguished awards, the Wright Brothers Memorial trophy, bestowed annually by the National Aeronautic Association through the Aero Club of Washington DC. In 1981, she was inducted into the Aviation Hall of Fame in Dayton, Ohio, joining her late husband who has been so honored in 1977. In 1983 Mrs. Beech was inducted into the National Business hall of Fame.

A Smattering of Scientists

Olga Taussky-Todd (1906 – 1995 CE) was a highly respected mathematician who studied algebraic number theory. She was born in Olmütz, now part of the Czech Republic. She received a D. Phil. from the University of Vienna in 1930, focusing on number theory. She is most closely identified with the field of linear algebra and matrix algebra where she made important contributions. During World War II she worked at the National

Physical Laboratory in Teddington, near London, investigating an aerodynamic phenomenon called flutter (an airplane when it flies too fast will become unstable, leading to flutter). The California Institute of Technology invited her and her husband to join the staff in 1957, he as a professor and she as a research associate. In 1971 she became a full professor.[8]

Maud Slye (1869 – 1954 CE) was known as the 'mouse lady' and sometimes called America's Marie Curie. She managed to produce highly inbred strains of mice in which nearly every specimen acquired a cancer. Even the type of cancer and the age at which it appeared could be predicted with reasonable and often astonishing accuracy. Dr. Slye advanced the view that cancer is due to a single genetic factor that determines nearly every aspect of the disease-the type of cancer, the site on which it arises, and the age at which it appears. Whether an individual acquired a cancer or not seemed to depend almost entirely upon his genetic endowment. This is one of the views about cancer development in humans popular in the 1920's that is no longer thought to be correct, but the specificity of genetic variants for development of tumors is well established. She published more than 40 booklets on cancer.

Inge Lehman (1888 – 1993 CE) was a Danish geophysicist who discovered the solid core of the Earth. This inner core has a diameter of about 1560 miles and is at a depth of about 3180 miles (Figure 65). She discovered this by examining thousands of earthquake records (she did this before computers). Some earthquake waves are refracted (P waves) as they pass through the inner Earth providing a map of the core.

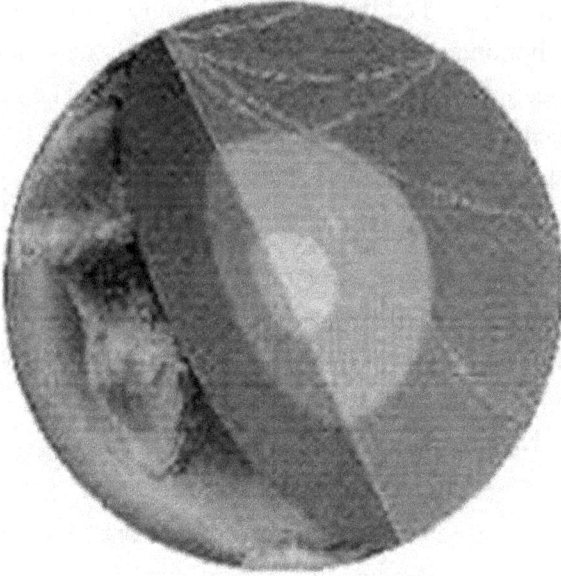

FIGURE 65 a cutaway of the Earth showing the core

Bessie (1891 – 1995 CE) and **Sadie Delaney** (1889 – 1999 CE) were two delightful ladies; born in the time of segregation they managed to obtain educations and reach out to others. Bessie became a teacher and Sadie a dentist. The book *Having Our Say* written by Sadie L. Delaney is the story of their lives and well worth reading.

Mary Ainsworth (1913 – 1999 CE) earned her PhD in developmental psychology in 1939 from the University of Toronto. She joined the Canadian Army during WWII. After the war her interests led her to Uganda where she was a Senior Research Fellow at the east Aafrican Institute for Social Research. She studied cultural differences in attachment formation in infants. The "strange situation" room she developed in which infants are placed during attachment testing is now a standard procedure. She settled at the University of Virginia in 1974 where she remained for the rest of her career. She wrote numerous books and articles and received many honors.

Rosalind Elsie Franklin (1920 – 1958 CE) was a chemist. She graduated from Newnham College (a women's college founded in 1871) at Cambridge in 1941. She was responsible for much of the research and discovery work that led to the understanding of the structure of deoxyribonucleic acid, DNA (Figure 66). The story of DNA is a tale of competition and

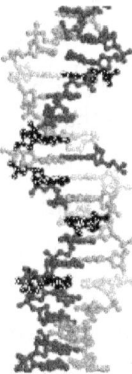

intrigue, told one way in James Watson's book *The Double Helix*, and quite another in Anne Sayre's study, *Rosalind Franklin and DNA*. James Watson, Francis Crick, and Maurice Wilkins received a Nobel Prize for the double-helix model of DNA in 1962, four years after Franklin's death at age 37 from ovarian cancer.

FIGURE 66 DNA structure

Some of these stories make it look rather bad for the women of the 20th century. But these women did succeed, and their stories are now being told.

Born Cecilia Payne (1900 – 1978 CE), she married Sergei Gaposchkin whom she rescued from Russia and was thereafter known as **Cecilia Payne Gaposchkin**. Her PhD dissertation was said to the best one in 20th century astronomy. She became the first woman to become a full professor at Harvard. She also was the first person to receive a PhD (in 1925) in astronomy from either Harvard or Radcliff. Her dissertation, entitled "Stellar Atmospheres, A Contribution to the Observational Study of High

Temperature in the Reversing Layers of Stars", folded the new field of quantum mechanics into spectroscopy and argued that the great variation in stellar absorption lines was due to differing amounts of ionization (differing temperatures), not different abundances of elements. She correctly posited that silicon, carbon, and other common metals seen in the Sun were found in about the same relative amounts as on Earth but the helium and particularly hydrogen were vastly more abundant (by about a factor of one million in the case of hydrogen). This result disagreed with earlier theories, and when she sent a draft of her paper to Dr. Henry Norris Russell, he replied that such a result was "clearly impossible." Russell had an earlier paper which argued that if the earth's crust were heated to the temperature of the sun the spectrum would look the same. Deferring to Russell's stature as an astronomer, Cecilia added the comment that her results were "almost certainly not real." Within a few short years most other astronomers had come to believe that hydrogen was far more abundant in the Sun than in Earth, and she was vindicated. In 1977 she received the prestigious Henry Norris Russell Prize from the American Astronomical Society. The following is an excerpt from her acceptance speech and memorial lecture for the Russell prize.

The reward of the young scientist is the emotional thrill of being the first person in the history of the world to see something or to understand something. Nothing can compare with that experience; it engenders what Thomas Huxley called the Divine Dipsomania. The reward of the old scientist is the sense of having seen a vague sketch grow into a masterly landscape. Not a finished picture, of course; a picture that is still growing in scope and detail with the application of new techniques and new skills. The old scientist cannot claim that the masterpiece is his own work. He may have roughed out part of the design, laid on a few strokes, but he has learned to accept the discoveries of others with the same delight that he experienced his own when he was young.

She was a colorful character in astronomy, known for smoking strong Russian cigarettes. I once saw her in the hallways of Harvard, striding down the hall, puffing on a cigarette, followed by eager students. Her autobiography *Cecilia Payne-Gaposchkin* was edited by her daughter and published by Cambridge University Press, 1984.

Dorritt Hoffleit (1907 – 2007) was an astronomer at Yale University. She completed her PhD at Radcliffe. She had Harlow Shapley (Director of Harvard College Observatory after Pickering) as her mentor. Still going strong at 100 years of age, she was honored by Yale at her centennial celebration. She said that during Shapley's regime there (1921 – 1952) fifty PhD degrees in astronomy were awarded by Harvard or its associated women's college, Radcliffe. The British Cecilia H. Payne (see above) was the first to fulfill the requirements. As Harvard steadfastly refused to award any degree to women, Radcliffe came to the rescue and awarded her the degree in 1925. The second under Shapley but the first awarded by Harvard was to the Canadian, Frank S. Hogg, in 1929. The third and fourth went to American women, to **Emma T. R. Williams** in 1930, and **Helen B. Sawyer Hogg** in 1931. The following year, 1932, an American woman, **Carol Anger**, and a Canadian, Peter M. Millman, earned the degree. Of the fifty doctorates awarded during Shapley's directorship, fourteen went to women. In 1956, Dorritt became director of the Maria Mitchell Observatory (see Chapter 5) and also held a position at Yale Observatory. She continued the tradition begun by Maria Mitchell by mentoring young women in astronomy. Always gracious in the face of adversity, her career spanned two world wars and the cold war. An interviewer once asked her what common and essential traits an astronomer must have in order to succeed. She answered

"The love of the subject, I think, is extremely important, not just the curiosity but the love of it. And then of course there is the curiosity and trying to satisfy the curiosity. How well these two characteristics go together, love and curiosity."

Dorothy Hill AC, CBE, FAA (1909 – 1998 CE) was a geologist and a graduate of the University of Queensland. She obtained a PhD from Cambridge University in 1932. The main area of work for her PhD was on the Carboniferous corals of Scotland. She was also a world authority on Paleozoic corals, and her publications remain the definitive work in the field to this day. She managed an impressive array of firsts:

- first woman to graduate with a gold medal
- first woman to be admitted as a Doctor of Science at the University of Queensland
- first professor who was a woman at an Australian university
- first woman elected as a Fellow of the Australian Academy of Science
- first president who was a woman, and the first woman Fellow of the Royal Society
- first member who was a woman of a Professorial (Academic) Board in Australia

Martha Chase (1928 – 2003 CE) was a biogeneticist known for her part in the pivotal 1952 "blender experiment" that firmly established DNA as the substance that transmits genetic information. Later she earned a PhD from the University of Southern California.

Maria Goeppert Mayer (1906 – 1972 CE) was a physicist who was not actively acknowledged by her peers until 1963 when she received the Nobel Prize in Physics for her work in the shell model of the nucleus (this work now appears in every basic physics text). She obtained her PhD in 1930 in theoretical physics at Göttingen.

This is another case of a woman denied, although she was finally recognized by her physics community and offered a professorship at the University of California, La Jolla. She was unable to take the position because of illness.

One might say that the 20th century does not sound very welcoming to women, especially the first half. In so many cases women had to struggle; they had that in common with their earlier sisters in science. By sheer stubbornness and cleverness they succeeded. Let me tell you a story about the 200 inch Palomar telescope (Figure 67).

FIGURE 67 Palomar Observatory

It was, at one time, the largest telescope in the world. Women were not allowed to observe with the telescope; not because they were incapable, but because the facilities had only one bathroom, and the men of California Institute of Technology who ran the telescope could not envision sharing a bathroom with a female! In fact on the request form for observing they put the phrase

It is not feasible for women to undertake an observing project.

Well, this was a challenge to Dr. **Vera Rubin** (astronomer) who in 1965 wrote an application to observe anyway. She took the form and carefully typed in the word 'usually' at the end.

It is not feasible for women to undertake an observing project, usually.

The review committee did not notice her gender or the change on the form and granted her the right to observe. When they figured out what they had done, the embarrassed group of men finally caved in to decency and granted her the observing time. When she showed up on the mountain to begin her observing run, the previous night's astronomer took her by the hand and led her to the bathroom, opened the door and said "feel free".

So you see women did get their way ... usually. Vera Rubin went on to a highly distinguished career in astronomy and eventually received the Presidential Medal of Science for her work.

Ruby Payne-Scott (1912 – 1981 CE) was the first woman to be a radio astronomer. In 1936 she finished a master's degree in physics from Sydney University in Australia (they did not offer PhDs at this time). She began her career in medical research in cancer radiology and then worked for the war effort during World

246

War II. In 1944 she with her colleagues hung an aerial 'horn' outside the lab window trying to detect cosmic static, and radio astronomy in Australia was born. She is credited along with her colleagues as being the first to relate solar radio emission to sunspots. Now we know that enormous solar flares can affect communication.

Maude Menton (c. 1913) working with Leonor Michaelis derived a mathematical model to describe the kinetics of how enzymes catalyze (enable) reactions. The work of Michaelis and German scientist Maude Menton enabled several subsequent generations of biochemists to assess correctly the nature and efficiency of the key, enzyme-driven steps in cell metabolism. Their joint work in 1913 led to the model describing the kinetics of how enzymes catalyze reactions, called the Michaelis–Menton equation (Figure 68).

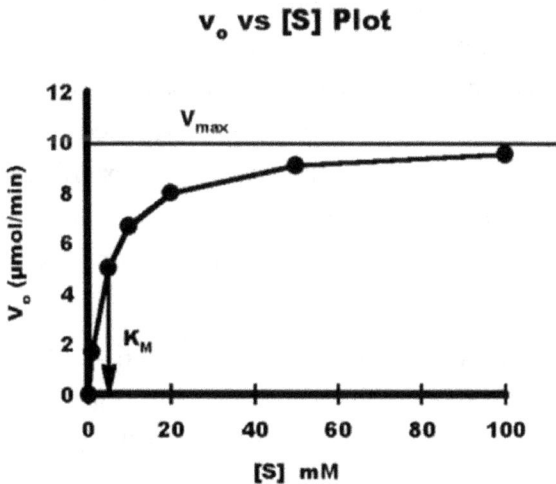

FIGURE 68 Michaelis-Menten saturation curve - V_o vs. [S] plot

Gertrude Scharff Goldhaber (1912 – 1998 CE) was a physicist who in 1948 confirmed the identity of beta rays and atomic electrons. She was one of the many scientists who fled Germany before World War II and came to London and eventually to the United States. While at the University of Illinois she discovered that neutrons are emitted in spontaneous fission (1942). She was the third woman, who was a physicist, elected to the National Academy of Sciences.

Lise Meitner was a German physicist (1878 – 1968 CE) who could be called the mother of the Atomic Bomb, a distinction she did not cherish. She along with Otto Hahn was the first to explain the splitting of the atom – atomic fission. Hahn received the Noble Prize for his part in this; she did not share the award. She published her study in *Nature* on January 16, 1939 thus ushering in the nuclear age.[9] She was the first woman admitted to the University of Vienna to study physics where she completed her PhD. She finally acquired an appointment in 1914, the first for a woman, in the University of Berlin physics department. She escaped from Germany when her life was in danger from Nazi restrictions and ended up in Sweden. She struggled for most of her life to obtain equality for women in science. A pacifist, she fought to maintain a distance from the atomic bomb weaponry made possible by her work with fission. She argued for peaceable use of atomic energy.

Grace Brewster Murray Hopper (1906 – 1992 CE) was a remarkable lady. She received a PhD from Yale University in 1934 in mathematics (Figure 69). She joined the US Navy where she remained for the rest of her career. She was the first woman to:

- Develop operating programs for the first automatically sequenced digital compute (1945)
- Develop the concept of automatic programming (1951) that lead to COBOL
- Receive the computer science Man of the Year award from the Data Processing Management Association (1969)
- Receive the US Medal of Technology (1991).

She was the oldest person on active duty in the US Navy when she retired at the age of eighty. She gave the most inspiring speeches. I was fortunate to hear her speak and receive one of her nanoseconds. She handed out telephone wires one nanosecond long (approximately 11.3 inches) to all the people attending her lectures. She often testified before the Congress, and said that one had to be explicit when talking to Congress, so she would use a nanosecond to explain why one has to wait for a reply from someone sending a message from the Moon. It takes a lot of nanoseconds lined up to reach the Moon.

She often said she invented the term computer bug and the log bears her out (Figure 70). The Mark I computer had died overnight, and they found a moth in the relay. The term 'bug' meaning defect in a machine, plan or the like was used long before this however. Thomas Edison is said to have discovered a "bug" in his phonograph, implying an imaginary insect. So although "computer bug" begins with Grace Hopper, the concept of 'bug' does not.

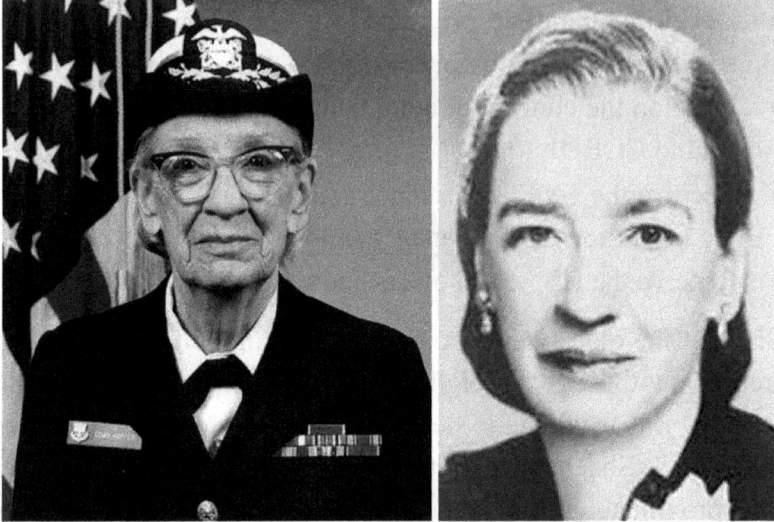

FIGURE 69

Photo # NH 96566-KN First Computer "Bug", 1945

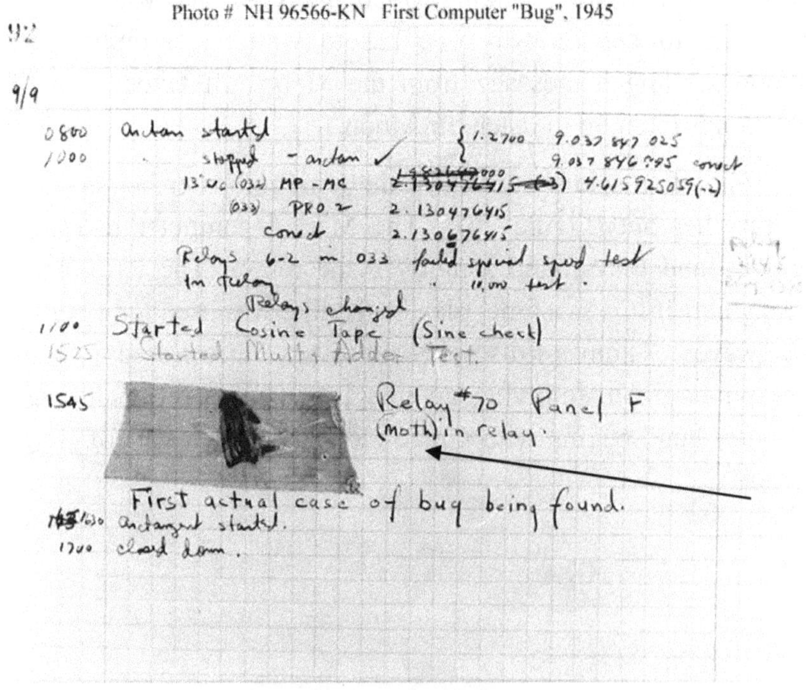

92

9/9

0800 Antan started
1000 stopped - antan ✓ {1.2700 9.037 847 025
9.037 846 795 correct
13"UC (033) MP-MC 2.130476415 (23) 4.615925059(-2)
(033) PRO 2 2.130476415
correct 2.130676415
Relays 6-2 in 033 failed special speed test
In relay 10,000 test.
Relays changed
1100 Started Cosine Tape (Sine check)
1525 Started Multi Adder Test.
1545 Relay #70 Panel F
(moth) in relay.
First actual case of bug being found.
1630 antangent started.
1700 closed down.

FIGURE 70

Frances "Betty" Snyder Holberton (died 2001) was another software pioneer who programmed that ENIAC computer. She also helped create the computer languages COBOL and FORTRAN.

Emma Markovna Trotskaya Lehmer (1906 – 2007 CE) was born in Russia. She studied at home until age 14, and then attended a community school. In 1924, she was accepted at UC Berkeley, where her favorite math teacher was Derrick Norman Lehmer, for whom she worked doing computation tables with his son Dick, a physics major who later became a mathematician. Four years later, she received her bachelor's degree with honors in math and married Dick. They had two children. The couple moved to Providence, R.I., where she received her master's degree in math in 1930. Her husband received his doctoral degree in math that same year. During the Great Depression, they migrated around the country from campus to campus – among them Cal Tech, Stanford, and Lehigh University in Pennsylvania – as Dick Lehmer assumed one teaching job after another. They also spent a year at the Institute for Advanced Study in Princeton, N.J. Emma balanced intellectual interests with motherhood: "It was a rare occasion when I could get a babysitter and come to somebody's lecture," she recalled in a 1984 interview. In 1940, Dick returned to UC Berkeley to accept a professorship in mathematics. Emma did war-related work in Berkeley's statistics lab in the 1940s and during the war years taught a few courses in statistics. Working as an independent scholar, she wrote or co-authored about 60 mathematical papers. She and her husband also worked on a famous, four-century-old mathematical problem known as Fermat's Last Theorem. She believed that mathematics had a sort of beauty that is irresistible to those who find it fascinating.

Charlotte Moore Sitterly (1899 – 1990 CE) was an astronomer who made key advances in our understanding of atomic structure. Astronomers and physicists could use both laboratory spectra to study specific transitions between energy levels of atoms or positive ions. Charlotte, already an expert in the compilation of atomic spectra and author of the Princeton Observatory *Multiplet Tables of Astrophysical Interest* of 1933 and 1945, accepted a position at the National Bureau of Standards in 1945 to prepare a handbook of atomic energy levels. This project achieved a first milestone in 1949 with the publication of Volume I of *Atomic Energy Levels,* containing the spectra of hydrogen through vanadium.

In 1952, Volume II with the elements chromium through niobium followed, and in 1958 a third volume containing the spectra of molybdenum through actinium completed this series, also known as NBS Circular 467. These are probably the preeminent resources for atomic spectroscopy data. Charlotte's atomic energy level books covered 75 chemical elements and 485 spectra in different ionization states. They became an essential tool for atomic, plasma, and astrophysicists as well as spectro-chemists. About 7000 copies of each book were sold, and the books were reprinted in 1971 as NSRDS-NBS 35. Charlotte did her PhD at the University of California, Berkeley, on the spectra of sunspots, then worked at Princeton University Observatory and produced there her first comprehensive spectroscopic compilation, *A Multiplet Table of Astrophysical Interest*, a first edition in 1933 and a revised and greatly enlarged one in 1945. Shortly afterwards she came to National Bureau of Standards (NBS)[10], where she was a member of the Atomic Spectroscopy Section until her retirement in 1968. After retirement, she continued her critical compilation work at NBS into her mid-eighties and also worked for several years at the Naval Research Laboratory on the ultraviolet spectrum of the sun. She exerted considerable influence on the field of spectroscopy for many years. She received numerous awards and honors, among them honorary PhD's from Swarthmore College, the University of Michigan, and the University of Kiel, Germany; the Department of Commerce Gold and Silver medals; and the William F. Meggers Award of the Optical Society of America. In 1961, she was one of six women who received the first Federal Woman's Award.

Lynn Margulis (born 1938) is the Distinguished University Professor in the Department of Geosciences at the University of Massachusetts, Amherst. She was elected to the National Academy of Sciences in 1983, received the Presidential Medal of Science in 1999. She works in cell biology and microbial evolution.

I could go on and on listing wonderful women of science and technology, and the list would reach around the Earth! Let these few stories live as gifts for the future. These women survived, they endured, and they triumphed because of that one singular moment, when they, suddenly, see something brand new. That moment is the rarest of joys, it becomes their existence. It makes all the struggles worthwhile. It defines their very being. After all, man does not define humanity. As Fanny Wright (1795 – 1852 CE) said

"Let women stand where they may, their position decides that of the race."

And the word scientist does not mean male. It is time to change the name! No longer are we 'women astronomers', but simply 'astronomers', no longer are we 'women scientists', but simply 'scientists'.

And in science we desperately need both points of view – male and female. When we, as scientists, wear blinders, about anything at all, we fail. I have never seen science succeed by using only one view, by using only one tool, by using only one person's thoughts, by looking at something only one way. These women succeeded. We need to celebrate them and consider them heroic. It is time to put our women of the past into our stories of the present and our hope for the future.

Despite the American standards provided by the National Academy of Sciences and National Academy of Engineering, a large percentage of high-school science and mathematics teachers lack an undergraduate or graduate major in a technical discipline or science education. Not only are they poorly prepared in the technical aspects of science and engineering, they are also ignorant of the history and social nature of science, mathematics, and engineering. What does this lack do? The student ill prepared ultimately affects us all negatively. Pope Pius II (1405 – 1464) offered this advice:

"You should support scholars as well as soldiers, for they will instruct you in making correct choices between wrong and right ..."[11]

Science and technology are important. To most of us high academic standards have become the last, best hope for saving America's schools. In the area of science and mathematics, student achievement as measured by the National Assessment of Educational Progress declined steadily from 1970 through the early 1980s from an already unacceptable level (National Center for Education Statistics 1997). The release in 1983 of *A Nation at Risk: The Imperative for Education Reform by the National Commission on Excellence in Education* warned of a national education crisis, and dozens of reports issued over the next few years supported the commission's conclusions and called for action. The reform landscape is crowded with projects, initiatives, centers, institutes, partnerships, and more. The most promising of these to emerge over the past decade or so share two common concerns: improving the quality of science and mathematics education and increasing the accessibility of science and mathematics education to students who had not participated previously.

Although things are improving, the notion that excellence is 'not for girls' (or minorities) persists. In 2010 the American Association of University Women issued a report *Why So Few* – a study of why there are still so few women in the STEM disciplines. Science, Technology, Engineering, Mathematics = STEM.

It is vital that teachers know what women have done, how they have contributed. Science and technology are *innately* diverse. We need role models that highlight and celebrate this diversity. We need to celebrate these women along with the men and raise them all to be heroes. Understanding science and technology will only strengthen our life, our work, and our world. We want solutions to our problems. They come from research, thought, and technology.

In addition, there is the wonderful news that by the end of the 20th century we have women by the thousands achieving advanced degrees in all the technical fields. It took 188 years for American women to get the vote; in the last 15 years American women earned over 15,000 PhDs in technical fields. Graduate schools in medicine and dentistry are routinely 50% female.

It is time to put our women of the past into our stories of the present and our hope for the future. Our hidden giants are no longer hidden. The pursuit of science is greater than any fantasy, than any game. Out of our joy in study and our endeavors on mountaintops, oceans, and labs come solutions to problems — the problems of the world. And we give it away freely — the best of gifts — the light of knowledge to our daughters and sons.

AS THE YEARS PASS

20th century   Valentina Tereshkova
Sally Ride
Lillian Moller Gilbreth
Edith Clarke
Vivien Kellems
Mabel MacFerran Rockwell
Elsie Eaves
Bette Nesmith Graham
Patsy Sherman
Twana Washington
Hedy Lamarr
Gladys Hobby
Elizabeth McCoy
Dorothy Fennel
Dorothy Hodgkin
Margaret Hutchinson
Anna Wessel Williams
Louise Pearce
Helen Brooke Taussig
Rachel F. Brown
Angela Ferguson
Gladys Dick
Mary Lura Sherrill
Lucy W. Pickett
Dorothy Wrinch
Gertrude Elion
Margaret Sanger
Georgia Dwelle
Rosalyn Sussman Yalow
Barbara McClintock
Margaret Meade
Rachel Carson
E. Lucy Braun
Marie Gertrude Rand

Donna Osif
Eleanor Raymond
Maria Telkes
Ida Eva Tacke
Chien Shiung Wu
Gerty Cori
Wanda Alma Marosz
Sylvia A. Earle
Julia Morgan
Evelyn Boyd Granville
Marjorie Lee Browne
Euphemia Loften-Hayes
Marguerite Thomas
Marie Daly
Bessie Coleman
Amelia Earhart
Mrs. Olive Ann Beech
Olga Taussky-Todd
Maud Slye
Inge Lehman
Bessie and Sadie Delaney
Mary Ainsworth
Rosalind Elsie Franklin
Cecilia Payne Gaposchkin
Dorritt Hoffleit
Emma T. R. Williams
Helen B. Sawyer Hogg
Carol Anger
Dorothy Hill
Martha Chase
Maria Goeppert Mayer
Vera Rubin
Maude Menton
Ruby Payne-Scott
Gertrude Scharff Goldhaber
Lise Meitner

Grace Brewster Murray Hopper
Frances "Betty" Snyder Holberton
Emma Markovna Trotskaya Lehmer
Charlotte Moore Sitterly
Lynn Margulis

---

[1] Although Sally Ride was the first American woman in space in 1983, NASA had trained 13 women who were pilots to become astronauts in the 1960's. They never did fly in space.

[2] Vivien Kellems was an exceptional engineer, businesswoman, and innovator. She provided the electrical and power engineering industries with newer and more efficient forms of grips and cables that revolutionized the industry. Her contributions to the World War II effort were invaluable.

[3] *The Book of Women's Firsts*, Read and Witlieb, Random House, 1992

[4] For further information on early women in chemistry of the northeastern US see *Early Women Chemists of the Northeast*, by N. Roscher and P. Ammons, *Journal of the Washington Academy of Sciences*, **71-4**, 1981.

[5] Ed. Kass-Simon and P. Farnes, Indiana University Press,1993

[6] The Comstock Act, 17 Stat. 598, enacted March 3, 1873, was a United States federal law which amended the Post Office Act and made it illegal to send any "obscene, lewd, and/or lascivious" materials through the mail, including contraceptive devices and information. In addition to banning contraceptives, this act also banned the distribution of information on abortion for educational purposes.

[7] http://www.math.buffalo.edu/mad/wmad0.html is the web site for Black Women in Mathematics.

[8] The web site for the Mathematical Association of America http://www.maa.org contains a nice biography of her.

[9] There is an excellent biography of her life: *Lise Meitner, Atomic Pioneer*, by D. Crawford, New York Crown Publications, 1969. Another biography is *Lise Meitner and the Dawn of the Nuclear Age* by Patricia Rife, 1999.

[10] Now called NIST – national institute of standards and technology

[11] *Renaissance Reader*, edited by Kenneth Atchity, Harper Collins, 1996, p.51

INDEX

5